Zoarces anguillaris,
Mugil.
Esocidae.
Labridae.
Lophidae.
Gobidae.
Siluridae.
Cyprinidae.
Salmonidae.
Clupidae.
Mugilidae.
Scombridae.
Chaetodontidae.
Sparidae.
Scienidae.
Triglidae. (Gurnards.)
Percidae. (Perch Family.)
Fins usually with some bony rays.
Salamandridae
Amphiumidae
Three Toed Amphiuma
Mud Devil, Ground Puppy, Hell bender.
Sirens. Eel like. No hind feet.
Proteus anguinus.
Siredon pisciformis
ABDOMINALES. Ventral fins behind the pectoral.
SUBBRANCHIALS.
APODES.
LOPHOBRANCHIA.
PLECTOGNATHI.
ACANTHOPTERIGII. (Spine rayed.)
MALACOPTERYGII. (Soft rayed.)
CTENOIDS. (Scales toothed on edge.)
CYCLOIDS. (Scales round.)
GANOIDS.
Orders of
Gymnodontidae.
The long tailed Unicorn fish.
Balistidae. 110 sp. (File Fishes.)
Balistes fuliginosus, Dusky Balistes.
Triqueter, or Dromedary Trunk fish,
Lactophrys Camelinus, Triangular body 3½ inches long.
Ostracionidae. 30 species. (Trunk fishes, 3 or 4 sided.)
Six horned Trunk fish, Ostracion sex cornutus.
Sauridae.
L. platyrhincus, Flat Nosed Bony Pike.
Lepisosteus bison, Buffalo, Bony Pike. Alligator Garfish, Garpike. (Scales enamelled.)
N. American, mostly fossil.
Polypterus of the Nile.
Without teeth, mouth beneath, branchiae (Gills) free.
ELEUTHEROPOMI. 1st Order.
Acipenser rubicundus, Lake Sturgeon, 3 or 4 ft. long.
Chimaera, Sea Ape.
Acipenser huso, Great Sturgeon. In the Caspian Sea, weighing 2500 lbs, 1,500,000 eggs in one fish.
Carcharias vulpes, Thresher Shark, Fox S., Swingle tail. (Gills fixed.)
Basking Shark, Selachus. 32 ft. long.
Squalidae. (Sharks. 114 sp.)
PLAGIOSTOMI. 2nd Order. (Transverse mouth)
PLACOIDS (Flat scales.)
Sturionidae. 24 sp.
25 ft. long.
Lamna, Porbeagle, Mackerel Shark.
Zygaena malleus, Hammerheaded Shark.
Mustellus canis, Houndfish, Spinax, spiny Dog fish.
Rock Shark, 2 to 4 feet long.
Scyllium canicula, Small spotted Dog fish, Rousselle, or Sea Dog.
Squatina angelus, Shark Ray, Monk fish.
Lophius Americanus, Sea Devil.
Pristis antiquorum, Sawfish, 18 or 20 feet long.
Raiidae. (Rays. 130 sp.)
CYCLOSTOMI. 3rd Order. (Round mouth)
BRANCHIOSTOMA. 4th Order.
Petromyzonidae. (14 sp.)
Amphioxus, Lancelet.
Sharks & Rays egg, ½ nat. size.
Pastinaca, Sting Ray.
Cephaloptera vampirus,
Raia clavata, Thornback,
Ammocoetes.
Libellula Dragonfly,
Honey Bee, Apis mellifica
Neuter, or Worker.
Male, or Drone
Privet Hawk-moth, Sphinx ligustri.
Tiger moth, Arctia caja.
Bombyx mori, Silk-worm moth in its last and perfect state called the Imago.
Telamona, Tree Hoppers.
Cicada septendecim
Lygaeus, Chinchbug.
Pentatoma.
Cimex lectularius, Bed Bug.
Reduvius, Skippers, or Water measurers.
Grand Father, Greybeard, Harry Longlegs.
Midge.
Tipula,
Octopus vulgaris, Poulpe,
Cuttle Fish, Polypus.
Eledone.
Argonaut, Argo, Argonaut, or Paper Nautilus.
Nautilus Pompilius, Pearl Nautilus.
Onychoteuthis
Common
Eggs.

HOUSE OF LOST WORLDS

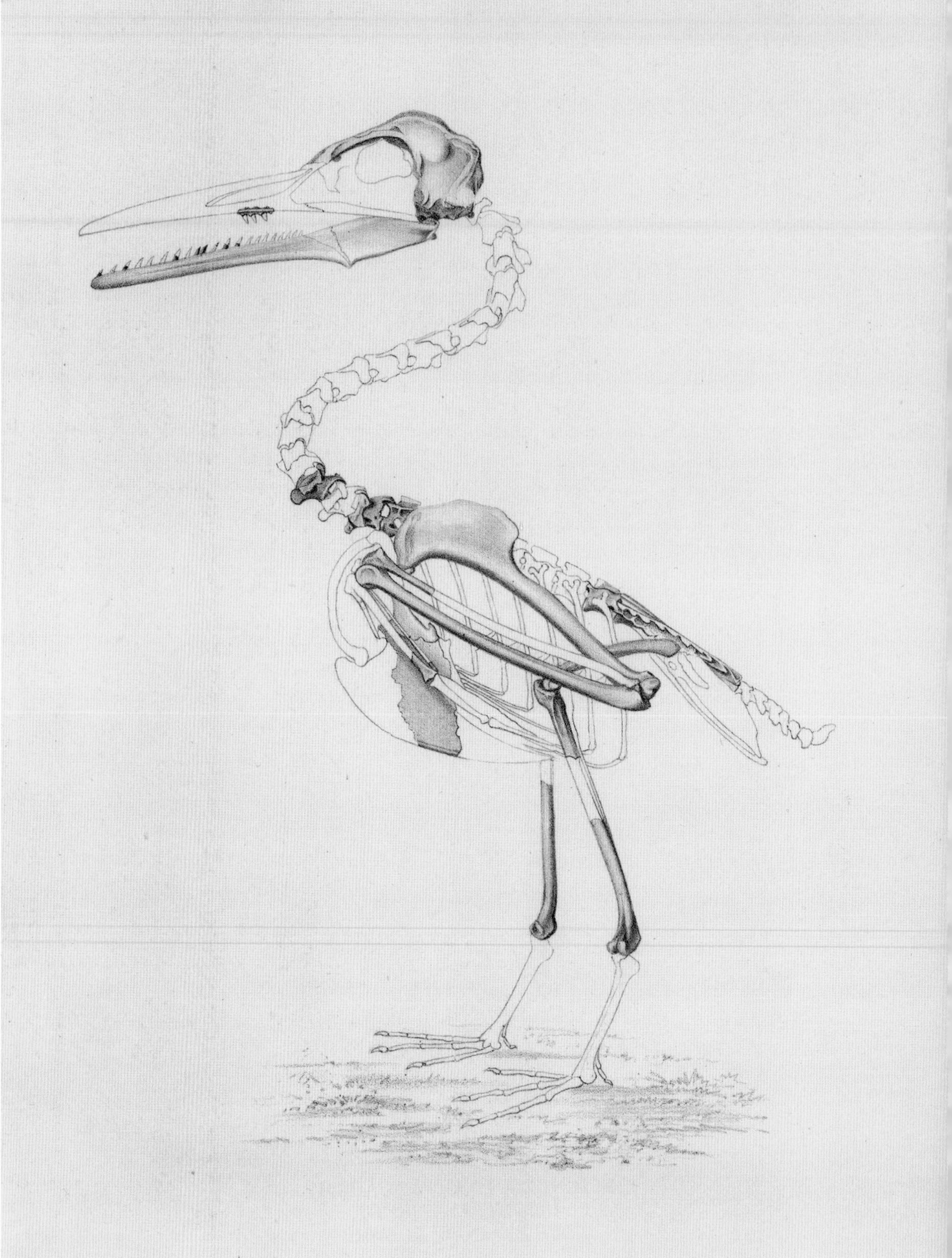

RICHARD CONNIFF

HOUSE *of* LOST WORLDS

Dinosaurs, Dynasties, & the Story of Life on Earth

Yale UNIVERSITY PRESS
NEW HAVEN AND LONDON

Published with assistance from the Fourth Century Trust.

Published with assistance from the foundation established in memory of Philip Hamilton McMillan of the Class of 1894, Yale College.

Yale University Press books may be purchased in quantity for educational, business, or promotional use. For information, please e-mail sales.press@yale.edu (U.S. office) or sales@yaleup.co.uk (U.K. office).

Designed by Nancy Ovedovitz.
Set in Arno Pro and Perpetua type by Tseng Information Systems, Inc.
Printed in China.

ISBN 978-0-300-21163-4 (hardback : alk. paper)

Library of Congress Control Number: 2015949553

A catalogue record for this book is available from the British Library.

This paper meets the requirements of ANSI/NISO Z39.48–1992 (Permanence of Paper).

10 9 8 7 6 5 4 3 2 1

Endpapers: A pre-Darwinian evolutionary tree published in 1858 by Anna Maria Redfield and purchased by Yale at the time as a wall chart.

Frontispiece, p. ii: *Ichthyornis dispar*, one of the toothed birds with which O. C. Marsh provided crucial early proof of evolutionary theory.

Page vi: A 500-million-year-old trilobite from Pennsylvania, one of the gems in the Peabody Museum's extensive collection of invertebrate fossils.

To the scientists of the

Yale Peabody Museum of Natural History

6657

CONTENTS

Introduction: Hunting for Truffles

Something about the word "museum" tends to make people feel very slightly dreary, but this is not a dreary museum and all museums, with my thinking, should be places of life and enjoyment and gaiety and fun because that is what education is all about.
S. DILLON RIPLEY, 1984, at the Peabody Museum

THE SCIENCE AND THE STRANGENESS BOTH begin in a parking lot on a busy street in New Haven, Connecticut, outside the Yale Peabody Museum of Natural History. The trees are honey locusts, an unspoken memorial to the mammoths and mastodons that once wandered there, consuming the tough, elongated pods of such trees and spreading their seeds. The vanity plate on the Subaru parked nearby, YPM 228, is the Peabody's catalogue number of a fossil trilobite, *Triarthrus eatoni,* which wriggled its many limbs 450 million years ago near what is now Rome, New York. (This is vanity of an esoteric variety.) The bumper sticker on another car advises, "Don't believe everything you think."

There was, as I was soon to learn, plenty more strangeness and science—150 years' worth—inside the walls of the Peabody, together with an abundance of thinking and rethinking about the world. One day, early in my research for this book, a curator proudly pointed out the collection of overeducated tapeworms extracted from incoming Yale freshmen at the beginning of the twentieth century. Also the vial of maggots from a murder victim wrapped in a rug and . . . But I changed the subject as quickly as possible to a lovely mobile hanging over a workbench, with a big luna moth the color of translucent jade and a morpho butterfly like a radiant blue sky, along with other moths and butterflies, all as if in flight. The insects were suspended with hairs donated by various staffers, the object being to determine whose hair was the least visible. The winner would gain the right to give up even more hair to make preserved insects airborne for an upcoming exhibit.

To be honest, I love this sort of thing (even the tapeworms). Robert Louis Stevenson once described his boyhood manner of reading as digging "blithely after a certain sort of incident, like a pig for truffles." In his hunger for bright images and bold action, the young Stevenson foolishly skipped past "eloquence and thought, character and conversation," which I think most adult readers would regard as the truffles. But the pig analogy strikes me as exactly right about the oinky, snuffling character of the writer's work, and for me the Peabody Museum was a mother lode of truffles.

One day, for instance, I came across a somewhat plaintive letter about luggage left behind with a U.S. Cavalry lieutenant named A. L. Varney at an outpost in the Wyoming Territory in 1872: "I am still holding the Fossil Mosasaurus for you," the lieutenant wrote to paleontologist O. C. Marsh, "but as I expect to be ordered from this station soon please advise me what to do with it."[1] This was unclaimed

(Overleaf) Preparator Hugh Gibb (shown here in about 1914) had the immense task of assembling many of Marsh's discoveries for public display, like this *Torosaurus* skull, and the *Pteranodon* mounted behind him.

baggage of a high order, a *Mosasaurus* being a prehistoric sea monster up to sixty feet in length. Fortunately for Lieutenant Varney, it was just the skull, now safely preserved as YPM.00327.

Another day, in the Historical Instruments Collection, I became fixated on a saliometer—that is, a saliva meter. It was an instrument I had not previously imagined to have existed. It was the one Ivan Pavlov used in his classic conditioning experiments on salivary response in dogs. (Pavlov gave it to Yale psychologist Robert Yerkes, from whom it made its way to the Peabody Museum.) In the Vertebrate Zoology Collection, I became enchanted with the story of the sled dog Togo, hero of a 1925 race to deliver diphtheria antitoxin through impossible weather conditions to a stricken Alaskan community. Togo lives on as the continuing inspiration for the Iditarod Trail Sled Dog Race—but his skeleton is now lovingly preserved in climate-controlled conditions at the Peabody Museum.

Truffle hunting, it turns out, has long been a familiar pursuit to the Peabody Museum staff. A vertebrate paleontologist there once remarked that the best collecting he had ever done was not in the Badlands of the American West nor in Inner Mongolia but in the basement of the Peabody Museum. The discoveries lie waiting even now—not just in the basement but almost anywhere in the cluster of buildings around the main museum as well as in archive and storage facilities in the suburb of West Haven. In 2002, for instance, researchers browsing through the paleontological specimens noticed a distinctive five-inch-long tooth. A Colorado collector had sent it in 1874 to O. C. Marsh, who was too swamped with other specimens to make sense of it. This was regrettable, as it has turned out to be the earliest known specimen of the über-predator *Tyrannosaurus rex* ("King of the Tyrant Lizards"). Thus Marsh, an intensely possessive and competitive collector, missed out on the discovery, and the species had to wait another thirty-one years to be described—by one of his hated rivals.

Somewhat more humbly, a curator browsing through the botanical collection a few years ago noticed a liverwort specimen marked as the type of its species—that is, the individual by which the species was first described and defined. It looked like a bunch of dried schmutz, as preserved liverworts will do. Then the botanist looked more closely at the identifying tag and found it had been collected by one Charles Darwin during his travels aboard HMS *Beagle.*

Over 150 years, the Peabody Museum has collected, described, catalogued, and preserved roughly 14 million specimens—representative samples of the Earth, its succession of inhabitants, and the heavens. Many Earths, really, over several billion years. A relative handful of these specimens are on public display. Many more wait for someone to come along, next year or 150 years from now, with the right

The Mystery of the "Intercostal Clavicle"

This isn't perhaps what the mercantile banker George Peabody had in mind when he founded his namesake museum in 1866, nor is it something to which the scientists working there have ever aspired. But the Peabody Museum has, on the side, enjoyed (or endured) a lively career in popular entertainment.

The 1938 screwball comedy *Bringing Up Baby,* for instance, featured Cary Grant as a mild-mannered paleontologist and Katharine Hepburn as a flighty heiress with a pet leopard named Baby. The plot turns on a *Brontosaurus* specimen a lot like the one that had recently been mounted at the Peabody Museum. The resemblance was the result of a Hollywood researcher's having come to the Peabody and climbed an eighteen-foot ladder to measure and photograph every bone. In the movie, the hero ends up hunting for a bone that never existed in real life, the "intercostal clavicle." More realistically, he is also desperately seeking a donation from a Mr. Peabody.

The Peabody Museum was also the inspiration for the archetypal king of movie monsters, Godzilla, though at second hand. The art director of the original 1954 Japanese production studied the Peabody Museum's influential *Age of Reptiles* mural, then a recent cover story in *Life* magazine, and concocted a creature based on artist Rudolph F. Zallinger's *Tyrannosaurus, Iguanodon,* and *Stegosaurus.* (Zallinger's mural also inspired a generation of future paleontologists, possibly with help from Godzilla.)

More recently, in the long-running cartoon series *The Simpsons,* Yale alumnus and megalomaniacal plutocrat Mr. Burns and a character named Lily Bancroft were revealed to have had sex at the Peabody Museum in a (fictional) diorama of Eskimos and penguins, begetting a child voiced by Rodney Dangerfield.

At this writing, HBO is planning a series based on the bitter nineteenth-century Bone Wars between archrival paleontologists O. C. Marsh of the Peabody Museum and Edward Drinker Cope of the Academy of Natural Sciences of Philadelphia. One Cope biographer responded to the report that it will be a comedy with a four-letter word: "GACK."[a]

blend of knowledge, insight, luck, or technology to make sense of them. They wait for someone to whom a drawerful of specimens, silent for ages, may begin to resonate with lost voices and ancient dreams.

"I pound away at my typewriter and squint industriously through my microscope," a paleontology graduate student once wrote to his sister as he studied fossils in the back rooms of the Peabody Museum. "I publish articles and give papers and meet people and try not to laugh, but the moon is always over my shoulder." Then, in four eloquent sentences, the writer, whose name was George Gaylord Simpson, explained the feelings that kept him at it: "The spectacle at which I attend is vastly moving. There is an almost painfully epic sweep to the vastness of geologic ages which pass beneath my fingers in tattered fragments. The commonplace room is always filled with the mute cries of ages impossible to contemplate in which life has blindly toiled upward, or at least to further complication and further ability to realize that it cannot realize anything at all. It is all very strange and thrilling in a way which is, I am afraid, incommunicable."[2]

The people doing this work are no doubt prone to the ordinary (but also engaging) human business of petty resentments and bickering. And yet the work they do, this business of looking at the same specimens over and over, pondering how two bones might once have articulated together or deciphering how a certain plant once lived, is magical. A jigsaw puzzle of broken rock somehow opens a path millions of years into the past. It has the power to resurrect species and re-create worlds. This is what makes the Peabody Museum so appealing for me as a writer—and, I hope, for readers as well—especially because the results from work there have often been so profoundly influential.

The Peabody has never been a particularly big museum (its staff currently numbers about eighty-five), and New Haven is not even the biggest city in Connecticut. (Among U.S. cities, it ranks 195th just behind Olathe, Kansas, and Sterling Heights, Michigan.) But it suggests something of the Peabody Museum's outsized influence that it was a major force in creating, at roughly the same moment, both the age of oil and the American conservation movement. It was the Peabody Museum that largely drove the first great age of dinosaur discovery in the nineteenth century. It was the Peabody, more recently, that killed off the image of dinosaurs as plodding and stupid, reinventing them as swift, agile, and even intelligent. It thus set loose a global pandemic of "dinomania," especially among schoolchildren, culminating in the multiple iterations of "Jurassic Park." Researchers from the Peabody Museum have been remaking the world for 150 years, from the introduction of science education in North America to the rise of modern ecology, from the first full description of the giant squid to the modern understanding of how feathers evolved, and from the uncovering of the monumental Inca ruins at Machu Picchu to the rise of Caribbean archaeology.

For me as a writer, the story of the Peabody has been irresistible. Two thoughts will, I think, help make it equally compelling to read. The first is a caveat: the scientists who managed these achievements were often brilliant scholars and otherwise admirable, even heroic, in the face of widespread mistrust of science. But until late in the twentieth century, they were also almost exclusively men, and they admitted into their ranks only other scholars who happened to be white, male, Protestant, and preferably members of established Yankee families, or, better still, related by blood or marriage. Of one such professor, whose father was also a professor at Yale, the historian Robert V. Bruce wrote that if he "had been born female, black, or a child of the nineteenth-century American South or frontier West"—or, for that matter, Irish, Asian, Hispanic, Jewish, or an immigrant of almost anywhere other than northern Europe—"it seems unlikely that his genius would have been heard. His social, economic, cultural, geographical, and institutional circumstances did not make him a genius," Bruce continued.[3] But they made his genius matter and allowed it to flourish. In honoring these men, we should bear in mind the many voices those same circumstances silenced.

There is, however, a tendency now to regard the story of "dead white men" as no longer worth telling and to exaggerate or romanticize the role of people who were ruthlessly excluded from power. To me it makes more sense to relate how history actually happened, bigotry and all, and this book has given me a very personal way to put this in perspective. In the course of my research, I discovered that my great-great-grandfather had been an immigrant laborer in a New Jersey marl pit at the time Marsh recovered a specimen of the sea monster *Mosasaurus* there. He was the sort of person Marsh would undoubtedly have regarded as beneath notice. And yet I am certain that my forebear, who later died in a collapse at that marl pit, had enough love of learning and of higher things to be proud to be associated with such discoveries. He would also understand that Marsh, not Conniff, is the better story.

The second thought that readers may find helpful is more like an outline of the book's organization. Five overarching stories or themes run through the book and sometimes overlap. Readers may be surprised, first of all, that for the ten opening chapters, the museum does not exist, at least not as a physical structure. But the dream of the museum motivated some of the greatest American scientists of the nineteenth century. James Dwight Dana in particular had spent four years circling the world and gathering specimens with the U.S. Exploring Expedition, at frequent risk of death, only to see, on his return home, many specimens lost to science because of bureaucratic bungling. He threw his heart and mind into making the Peabody Museum a place where specimens would be preserved as a permanent means of interpreting and understanding the world.

The grandiose personality of O. C. Marsh and the size and strangeness of the fossil animals he discovered dominate much of the book. A few of his collectors also appear here, notably the poker-playing, gun-toting visionary reader of rock, John Bell Hatcher. Marsh's nakedly personal combat with archrival Edward Drinker Cope runs through many chapters, culminating in the very public "Bone Wars" of the 1890s.

Though it often seemed like one man's museum in this era, the Peabody was always more than O. C. Marsh and more than dinosaurs. As the book progresses, I aim to bring this less public museum into the foreground. I confess I initially underestimated the scientists who worked in Marsh's shadow, but I soon found such lesser-known figures as Addison Emery Verrill, George Bird Grinnell, Charles E. Beecher, and Charles Schuchert walking around in my head, highly engaging characters whose contributions to both science and conservation still resonate today. If you are looking at the latest studies of the life around Atlantic seamounts, for instance, you should know that Verrill got there first.

The twentieth century opens with Yale's astonishing 1905 decision to demolish the museum building in the aftermath of some of the greatest scientific discoveries of the nineteenth century. Readers cannot help noticing here, as throughout the book, a paradox: for much of its history, major achievements by the Peabody Museum failed to elicit commensurate support from Yale. This was, I think, partly a product of an academic culture that often glorified intellectual theorizing above the object being theorized about. As an English major, for instance, I often had the impression that literary criticism mattered more than literature. This tendency seemed to be exaggerated when the objects came from the natural world rather than from the human mind. Until the 1950s, and even beyond, Yale also regarded the sciences as a second-class intellectual enterprise—much too hands-on, much too how-to—and the Peabody Museum perhaps more so than the rest. Science taught lessons about humanity that the humanities did not necessarily want to hear. The book also describes the eventual construction of a new museum building after nine years of homelessness and the struggle to figure out how a natural history museum should engage the attention of the public (not to mention the university). The courageous response, just as the 1925 Scopes trial was criminalizing the teaching of evolution, was to organize the Peabody Museum displays entirely on evolutionary lines and to teach evolution to schoolchildren and adults alike.

Toward the end of the book, I look at two major areas of research that struggled to get equal time or space with paleontology in the Peabody Museum. Anthropology once again put the museum in national headlines when Hiram Bingham came back with his account of the "lost city" of Machu Picchu. George G. MacCurdy and, later, a post–World War II generation of his intellectual grandchildren proceeded to develop anthropology as a serious science. Zoology went through a similar twentieth-century transition, with S. Dillon Ripley and Charles Remington building up major collections in their fields, and G. Evelyn Hutchinson and his students developing the theoretical framework of modern ecology. The book ends, however, as it began, back in paleontology, the study of lost worlds being the Peabody Museum's great strength.

In telling this story, my ambition is not just to explain the science but to let readers experience it for themselves. In that spirit, the place to start is back on the footpath that runs from the parking lot around the front of the Peabody Museum. It's embedded with Big Bird footprints (technically from a theropod dinosaur that walked 110 million years ago in Texas). The path runs through a Cretaceous garden of plants that dinosaurs knew—Allegheny spurge, dwarf mountain pine, cinnamon ferns, and tulip trees—and it loops out and around a model of a huge *Torosaurus* with gaping jaws, as if waiting to pluck up small children. (No worries, *Torosaurus* was a vegetarian. *Deinonychus,* on the other hand . . .) It leads the visitor to the entry door and into a lobby where a giant squid wriggles overhead and massive dinosaurs wait to be admired just around the corner.

Please step inside. You may find some truffles. You may hear the mute cry of ages. You may, with luck, end up no longer believing the things you thought when you came in.

CHAPTER 1

A West That Was Actually Wild

We stood upon the brink of a vast basin, so desolate, wild, and broken, so lifeless and silent, that it seemed like the ruins of a lost world.

CHARLES W. BETTS, "The Yale College Expedition of 1870"

On the last day of June 1870, a party of young men, together with their professor, left New Haven, Connecticut, bound for what they called the "Far West." One of them, a senior who had celebrated the end of his college years too ardently, appeared on the station platform only after the 9:30 a.m. train had already begun to move. "Hold the train!" cried the friends who had hustled him to the station. "Passenger for China!" one of them added, and the words "so startled the brakeman that he pulled the cord."[1] Their true destination was no farther than California, but that seemed far enough. They intended to travel the transcontinental railroad, completed only the year before, and strike out at various points north or south of the line to hunt for fossils. They were students and newly minted graduates of Yale.

The territories they would visit were largely unmapped and sometimes hostile. To anyone other than "the devoted and courageous student of science," Harry D. Ziegler, a member of the party, wrote upon returning to the East late that year, the journey "wore an aspect forbidding in the extreme." Hence "the greatest care had to be exercised in selecting the gentlemen who should compose the expedition.... Every member was chosen on account of his mental and physical fitness as well as because of his scientific attainments."[2]

Or perhaps not quite, thought George Bird Grinnell, also a member of the party: "None of us knew or cared anything about the objects for which [the expedition] was being undertaken," he admitted in old age, at the end of a distinguished career as one of the nation's pioneering conservationists. Grinnell had successfully nominated many of his own pals from the graduating class for the expedition, he recalled, based largely on friendship and a spirit of adventure. "Vertebrate fossils meant nothing to us, but we all longed to get out into the uninhabited and then unknown West, to shoot buffalo and to fight Indians."[3] (These youthful ambitions notwithstanding, Grinnell would go on to become one of the great saviors of the buffalo and a friend of the Pawnees, Blackfeet, and Cheyenne.)

Ziegler cast the party in the heroic mold: the men were "armed with a brace of trusty pistols, and the all-essential knife, as well as the geological hammer, and the appliances of science." He thought they were "not unlike" Julius Caesar when he led the Roman armies into Gaul, and the Yale men, too, "went, saw and conquered."[4]

Or rather, they were the rawest sorts of amateurs, said Grinnell, "young men from the East, some of whom had never mounted a horse," equipped with rifles they fired for the first time only upon crossing the Missouri River.[5] Many of them were scions of wealth and privilege. The party included a grandson and namesake of the cotton gin inventor, Eli Whitney, as well as the sons of a prominent midwestern mercantile banker, a manufacturer who had backed the construction of the Civil War ironclad USS *Monitor,* and a general who had died at the battle of the Wilderness. They had to pay many of their own expenses, and one of them later suspected that they might have been chosen largely on the basis of "parentals apparently commanding the needful." In Chicago, they posed for a group photograph, casually self-possessed in their city gear, from their bow ties and watch fobs down to their polished boots. (Some of them also undertook a challenge to have a drink in every saloon on a three-block crawl to their hotel.) "It was an entirely innocent party of 'pilgrims,' starting out to face dangers of which they were wholly ignorant," Grinnell wrote. "At this time the Sioux and Cheyenne Indians occupied the country of western Nebraska and that to the north and northwest, and they objected strongly to the passage of people through their territory, and when they could do so—that is, when they believed they had the advantage,—attacked such parties."[6]

After the men had traveled stop-and-start for two weeks, their Union Pacific train came to a halt one night in the middle of nowhere, at the shabby little frontier settlement of North Platte in western Nebraska. It deposited them beside the tracks and drew away into the encompassing darkness. Having heard "disquieting" rumors of "trouble ahead," the men had buckled on their revolvers a half hour earlier. One of them later recalled that an unfamiliar "earnestness and absence of frivolity" marked the occasion. They were still eight miles from their immediate destination of Fort McPherson, the intended starting point for their first fossil-

O. C. Marsh (*standing, center*) and his crew of urbane but inexperienced Yale men posed in Chicago for a final photograph in July 1870 before the first Yale College scientific expedition headed into the unknown West.

hunting trip. The expedition leader, their professor, had gone ahead a few days earlier to make arrangements, and a military wagon now waited to carry them onward in the dark, over rough roads, past bluffs and gullies that seemed like "excellent lurking places for the Indians," Grinnell recalled.[7] An army officer pointed out, purely as a matter of interest, a ranch the Indians had raided the previous fall—scalping all inhabitants.

Silent, and with spirits low, the students eventually arrived at the fort and presumably slept. They woke to learn that Indians "had swooped down" on a party of hunters nearby. One hunter, hit by an arrow, managed to fire at a retreating warrior, dropping him from his horse. A buckskin-clad guide had led a small party out to investigate and returned with the dead warrior's moccasins, Grinnell wrote, "at which the newcomers from New Haven stared in wonder." The Indian was a boy, a bit younger than themselves, and the guide wasn't much older. His name was William F. Cody, twenty-four years old and already becoming famous as Buffalo Bill. The swaggering Yale men had tumbled off the train directly into a West, as Grinnell put it, "that was then actually wild."[8]

For Marsh's crew, the "humming" of rattlesnakes "soon became an old tune," wrote Charles Betts, "and the charm of shooting the wretches wore away"—except after a rare chance to bathe, when rattlers chose to bask amid their shed clothing.

Professor M

The expedition's leader at least was a sort of Caesar. Othniel Charles Marsh, thirty-eight years old, was a skilled shot and an avid outdoorsman, thickening now into middle age. In photographs of the party, Marsh stands at the center, stolid and strong, with a broad high forehead, pale, wide-set eyes staring straight into the camera, a prominent nose, and a frizzled beard fanning out across his chin. He preferred to be known simply as "O. C.," and his students generally referred to him as "Professor M" or "the professor."[9] But that peculiar forename Othniel, the "Lion of God" in the Old Testament, hinted accurately at his ferocity and determination. Four years earlier at Yale, he had become North America's first professor of paleontology, and he intended not just to conquer this burgeoning field of science but to possess it entirely as his personal empire. It would require substantial institutional support. So Marsh had already set out to create the museum that would be the chief instrument and ultimate beneficiary of this vast ambition. The museum would be the gift of his uncle George Peabody, a London-based merchant banker, and Marsh would use it to achieve extraordinary things, transforming our ideas about the history of the North American continent and about life on Earth.

Marsh's 1870 expedition was merely the opening move in a much larger campaign to establish himself and his museum in the scientific world. In preparation, he had penciled terse and not particularly grammatical reminders in a pocket notebook. "Get Transcontinental RR Guide for 1870 Chicago George A Crofut 75 cents," he urged himself early on and later, more specifically: "Two days N fr Ft McPherson would reach . . . best bone region. Probably South Branch of Loup Cr would be good loc. Also Dismal River See Haydens map." Other notes had to do with people who might be helpful en route: "Jackson Abney Cheyenne Knows loc & will collect."[10]

Here at Fort McPherson and at other stops on the itinerary, the Yale party would travel under the watchful protection of a U.S. Army patrol. (It was not strictly speaking legal for the army to provide this service, but Marsh had made arrangements with Generals William Tecumseh Sherman and Philip Sheridan to dispatch patrols on routes that happened, by the purest coincidence, to be identical with those of the Yale party.) As the Yale men prepared to depart for the first time into the wilderness, U.S. Army major Frank North, a well-known scout and Indian fighter, eyed them skeptically. Then he took them out to the corral to introduce them to "a lot of Indian ponies captured from the Cheyennes" in a recent fight. North and two of his Pawnee Indian scouts "chose for us the gentlest mounts," Grinnell wrote, and then led them out into the sand hills.[11]

The soldiers were curious to know what this peculiar mission was all about and, on the first day's ride twelve miles north, Marsh told them, in grand professorial style. He was, as a student recalled it, "fairly on the war path" and "all aglow with achievements his imagination pictured as in store for him."[12] Marsh had made a brief exploratory visit hereabouts in 1868, just two years earlier, and no doubt he told them about it. "It was my first visit to the far West, and all was new and strange. I had a general idea of the geological features of the country I was to pass through" but no conception of the endless expanse of the Great Plains, "as far as the eye could reach in every direction." It reminded him "of mid ocean with its long rolling waves brought suddenly to rest." He explained that the grassy landscape all around them had in eons past been "a great fresh-water lake." Then he described the astonishing animals, now extinct, that they would find buried along its former shoreline.[13]

He had made that 1868 foray to run down a curious newspaper report. A railroad construction worker digging a well at Antelope Station (now Kimball, Nebraska) had unearthed ancient bones, said by a local doctor to be the remains of early humans. With his characteristic knack for making useful friends, Marsh had persuaded a conductor to hold the train there long enough for him to conduct a quick survey of the well site. He recognized several species, hominids not among them, before the conductor finally hurried him along. On the return trip the stationmaster, having been promised a reward, handed Marsh a hatful of other bones from the well. From this dubious treasure, Marsh had gone on to identify at least eleven different species from millions of years in the past. "I was not long in deciding," Marsh declared of the American West, "that its past history and all connected with it would form a new study in geology, worthy of a student's best work, even if it required the labor of a lifetime."[14]

Camels had once walked here, Marsh could now tell his military escort, and horses that were perhaps ancestors of the very ones on which they rode—but "scarcely a yard in height" and with three toes instead of a hoof. If so many discoveries could come from a single well, he noted, "what untold treasures must there be in the whole Rocky Mountain region." And thus, Marsh declared, "my own life work seemed laid out before me."[15]

The soldiers surely thought they had fallen into the hands of a madman. Grinnell later recaptured their feelings in his western novel *Jack among the Indians,* in which a character recollects "one of those professors we see out in the country sometimes; them fellows that come out to dig bones . . . I heard one of them talking once, and he just kept on for two or three hours, telling us about how the earth was made, and how this used to be water where it is all dry now, and a whole parcel of things that I didn't understand, and I don't believe anybody else did, except the man that was talking."[16]

Buffalo Bill Cody, who would become one of the great yarn spinners of the American West, was riding with the Yale party that first day, out of curiosity. He was unmistakably impressed by Marsh, perhaps most of all by this easterner's knack for the tall tale. "The professor told the boys some mighty tough yarns today," Cody remarked to the soldiers around the campfire that night. "But he tipped me a wink, as much as to say, 'You know how it is yourself, Bill!'"[17]

Into Lost Worlds

In fact, neither of them had any idea how it was, and their tallest tales could not come close to the reality of the discoveries that Marsh would in time bring to light in the western landscape. In the decades that followed, Marsh would go on to discover or describe — and bring back to the Peabody — new species by the railcar load. Among them, for instance, was a five-ton Jurassic monster with a double row of plates down its spine and spikes on its tail, now familiar to every schoolchild as *Stegosaurus.* Another, seventy-five feet long with a whiplike tail, soon came to life in the public imagination as *Brontosaurus,* "the thunder lizard." "As the exotic shapes of prehistoric beasts began to be assembled," first in print and later "for school children in Yale's Peabody Museum, it was clear that [Marsh] had discovered a new Western horizon," the historian William H. Goetzmann wrote. "He had in fact revealed and dramatized much of ancient America."[18]

As a result, the Peabody Museum would house some of the best early evidence linking dinosaurs to modern birds through what Marsh called "the discovery of birdlike Reptiles and reptilian Birds." (The full meaning of that link would become evident only a century later through other research at the Peabody.) It would also hold a series of discoveries by Marsh revealing the detailed evolution of the horse from its origins in the New World. Those collections, together with Marsh's personal convictions, would make the Peabody one of the nation's earliest and most enthusiastic exponents of Darwinian evolution. In an 1880 letter, Charles Darwin himself would remark that Marsh's work had "afforded the best support to the theory of evolution" in twenty years — that is, in the entire period since Darwin had first fully expounded the idea in his 1859 book *On the Origin of Species.*[19]

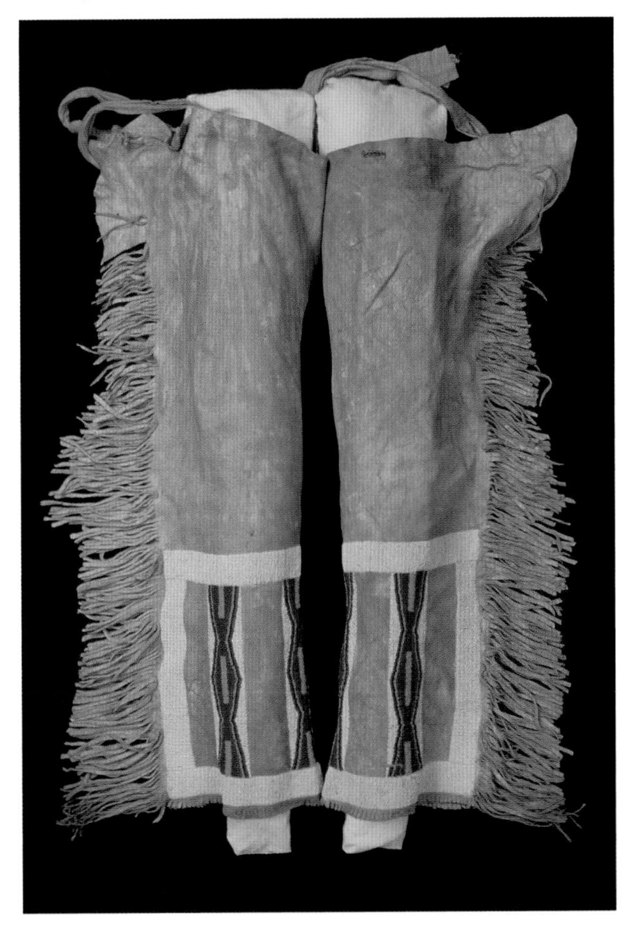

The expeditions were shadowed by different tribes and guided by others. These leggings were a gift to George Bird Grinnell from the Piegan Blackfeet.

Marsh's expedition and its sequels

"had a lasting importance not only in terms of his startling discoveries," Goetzmann wrote, "but also in terms of the effect it had on the future course of science and public policy in the West, and on the United States as a whole as Marsh rose to prominence in the worlds of science and government." It was the end of the era "of the pure military reconnaissance. . . . The emphasis had shifted to a more sophisticated approach to Western exploration."[20]

Marsh, who craved public attention and esteem, would become one of the nation's most famous scientists. Mark Twain would satirize him indirectly as "Professor Bull Frog" in an 1875 story about a hapless scientific expedition in the American West. A Kansas humorist would dub him "Professor Paleozoic." More seriously, Marsh's reputation, ambition, and influence would eventually entangle him in the infamous "Bone Wars" with paleontologist Edward Drinker Cope from the Academy of Natural Sciences of Philadelphia. The two men would become "the most famous haters in the history of science," as one account of that brutal rivalry put it, and the feud would be a nightmare for them both, bordering on tragedy, the more so because it played as comedy to the general public. It would drive each of them to an early and solitary death. But it would also make the two combatants "the central figures in the heroic age of American vertebrate paleontology," to the everlasting benefit of the museums on whose behalf they collected.[21]

All that was yet to come, though. As Marsh and his party headed out from Fort McPherson in July 1870, the museum for which he hoped to pile up riches still existed mainly on paper. It would not have a building, much less doors with which to open it, until 1876. The Peabody Museum of Natural History existed, that is, mainly in O. C. Marsh's own prodigious imagination. But the dream of such a museum had tantalized scientific men at Yale for decades.

CHAPTER 2

The Patriarch

What laws had its action controlled
Where none its dark pathway could trace?
Or how long, all unseen, had it rolled
Through the depths of etherial space?
EDMUND TURNEY, "The Meteor"

On October 26, 1802, Benjamin Silliman, a twenty-two-year-old Yale-educated lawyer, boarded a stagecoach in New Haven for the long, dusty, motion sickness–inducing trip to Philadelphia. Under his arm, he carried a wooden candle box full of mineral specimens to be properly identified. They were remnants of a higgledy-piggledy museum of curiosities Yale had maintained for a time but then largely misplaced, not altogether regrettably. The original collection had included "moose antlers weighing 50 pounds . . . a chain nine feet long made by a blind man from a single stick of wood," and "a two-headed calf killed in Hamden."[1] It also included miscellaneous unlabeled mineral specimens. The candle box in which Silliman carried them to Philadelphia would enter Yale legend as the beginning of proper scientific collecting at the college. It was also the beginning of the collections that would later become the Peabody Museum of Natural History and of Yale's rise from a college to a university.

The polymath Thomas Jefferson was in the White House. But for most Americans then, science was still a foreign enterprise, somewhat nervously regarded. There were signs of growing interest in this strange idea of knowing the world not just by faith but by experiments, expeditions, and observation. But many people considered it a threat to their religion and to the idea of a classical education.

Timothy Dwight, then president of Yale (and sometimes called "the elder"

Benjamin Silliman Sr., shown in a miniature painted on ivory in about 1815, became the great promoter of science education in the United States.

to distinguish him from a grandson), saw that it was time for the college to branch out from its primary function as a training ground for Congregational ministers. With the rising interest in science, he thought it would be wise to add a faculty position in "chymistry and natural history." But Dwight was also a pastor and an adamant defender of the church's privileged standing in Connecticut. He said he could find no American who was qualified for the job, and he feared that "a foreigner, with his peculiar habits and prejudices, would not feel and act in unison with us . . . however able he might be in point of science."[2]

Instead, Dwight turned in 1801 to a recent Yale graduate and family friend who was known to be devout. So devout, in truth, that he could contentedly describe Yale as "a little temple" where "prayer and praise seem to be the delight of the greater part of the students." Benjamin Silliman admitted to being "startled and almost oppressed" by Dwight's job offer. He knew nothing about science. It was an inauspicious start.[3]

Dwight argued his way around Silliman's concerns. He pointed out that there were already plenty of lawyers, but as a scientist "the field will be all yours." He also advised Silliman not to pursue a job possibility he had been considering in Georgia because of the morally repugnant association with slavery. Dwight was an abolitionist, and at Yale Silliman had also become adamantly opposed to "the sin and shame of slavery." But his Fairfield County family was neck deep in the Connecticut practice of slaveholding. Dwight cleverly made the Georgia job sound like a kind of falling back.[4]

Thus Silliman soon found himself en route to Philadelphia to take a crash course in chemistry at the University of Pennsylvania, which had the nation's first school of medicine. Dwight, it turned out, had chosen wisely. Silliman was certainly devout. But after another year attending lectures in London and Edinburgh, he was also a knowledgeable and enthusiastic teacher of chemistry, and later of mineralogy and geology. He came back to lecture brilliantly at Yale, and these lectures were the beginning of science education in America, that is, of science for its own sake, not merely as an adjunct to medicine. Silliman would become a great name. He was "the Patriarch" of American science, according to Louis Agassiz,[5] the Swiss biologist who would take up the mantle of science education at Harvard in 1847. But he would achieve this status without making any great discoveries, without introducing any bold new concepts or systems, and without ever fitting the stereotype of the scientist as brooding genius.

In 1734, Yale became the first American college to obtain a compound microscope, from Culpeper in London. It's now part of the Peabody's Historical Scientific Instruments collection.

On the contrary, what American science needed then was "an organizer, a promoter, a teacher, a preacher, a public relations man, a communicator and coordinator, and an exemplar of professionalism," historian Robert V. Bruce wrote. "Silliman was all of these." He was a charismatic figure with a clear and forceful way of speaking and an impressive, even aristocratic physical presence — one former student recalled him as tall and lean, "erect as a general on parade and with a general's expression of great power," with a high brow, deep-set eyes, a thin straight nose, and slightly pursed lips — altogether inspiring confidence and even belief in his listeners.[6]

Silliman made it his mission to develop science and science education, first at Yale and later nationwide. He came to this work with that inglorious but essential trait that Bruce described as "effectiveness in procuring facilities and supplies." It wasn't just that he had a keen eye for new material to embellish the Yale collection; he was also adroit at wheedling funds out of the Yale Corporation to pay for these acquisitions. Much of this effort went in support of mineralogy, a topic of popular interest then. It afforded "a pleasant subject for scientific research," according to an 1816 account, and tended "to increase individual wealth" and "to improve and multiply arts and manufactures and thus promote the public good." Mineralogy also tantalized collectors with a kind of Christmas morning enchantment in the moment the geological hammer broke open the dull crust of weathered and mineralized stone to reveal jewel-like colors and shapes within.[7]

Mineralogy attracted personalities who were no less colorful. Silliman handed over $1,000 for one mineral collection, put together with profits from the most notorious quack medical invention of that era, Elisha Perkins's "Metallic Tractors."[8] Silliman's brother, a lawyer in Newport, arranged the purchase of another mineral collection brought from England by a doctor who then had the misfortune to die in a duel over his "too great familiarity" with the wife of a South Carolina plantation owner.[9]

One coveted acquisition eluded Silliman, at least at first. In the darkness before dawn on December 14, 1807, a "globe of fire," seemingly half the size of the full moon, blazed across the skies of western Connecticut. Darkened rooms went bright as day. Farmers started up from their chores or sat upright in bed in terror, afraid the Judgment Day had come. Three "loud and distinct reports" like cannon shots burst over the town of Weston, followed by "a continued rumbling, like that of a cannon-ball rolling over a floor, sometimes louder, and at other times fainter." It was a meteorite, estimated by Silliman to have been "not much less than 300 feet in diameter" before it broke apart and rained down in pieces.[10]

The Cabinet

A few days later, Silliman and a faculty friend, James L. Kingsley, arrived on the scene to gather eyewitness accounts and to obtain a few fragments by purchase (the weeping and gnashing of local farmers having given way to joyful profiteering). A detailed report by the two professors, including Silliman's chemical analysis, concluded that the meteorite had come from outer space. Their account was soon being read aloud before learned societies in Philadelphia, London, and Paris. President Thomas Jefferson was skeptical, supposedly remarking that it was "easier to believe that two Yankee Professors could lie than to admit that stones could fall from heaven." He had reason to mistrust Yale and Connecticut, both then bastions of anti-Jeffersonian Federalism. But the skepticism was real, even if the quote was probably apocryphal (see sidebar). In a letter two months later, Jefferson wondered pointedly how the meteorite "got into the clouds from whence it is supposed to have fallen."[11]

Like much of the educated world then, Jefferson was still struggling with the dogma-shattering idea, introduced just a dozen years earlier by the French comparative anatomist Georges Cuvier, that species created by God could become extinct. As a passionate advocate of scientific discovery and president of the American Philosophical Society in Philadelphia, Jefferson had authorized funds to excavate the bones of a mastodon—the very creature that alerted Cuvier to the possibility of extinction. And yet he also ardently pursued the hope that these mammoth creatures still lived in the unknown American West. Jefferson may have been equally torn now by the idea that the Earth God had made for man could be randomly bombarded from the heavens. (Though if it must be so, why not Connecticut?) These doubts were characteristic of a time in which ancient cultural and religious beliefs were constantly crashing up against new facts, and the United States in particular was struggling to come to terms with science.

In the course of their research, Silliman and Kingsley had spent several hours searching fruitlessly for one unusually large stone that had landed just across the Weston border in the town of Trumbull (Silliman's birthplace, as it happened). When it finally turned up after they'd gone back to New Haven, it weighed in at 36.5 pounds—and the lucky farmer who found it thought it was worth $500. An amateur mineral enthusiast named Colonel George Gibbs (the rank was honorific) placed the high bid. He was the heir to a Newport shipping fortune, which he seems to have had no great interest in preserving. Among other acquisitions, he had recently purchased and brought home the extensive mineral collection of a Russian count and another collection accumulated over forty years by a great patron of science in France, Jean Baptiste François Gigot d'Orcy, a tax collector for Louis XVI before the French Revolution.[12]

Even before the Weston meteorite episode, Silliman's brother in Newport had tipped him off about Gibbs. Silliman had even wangled permission from Gibbs's sister to examine part of the Gigot d'Orcy collection while Gibbs was traveling abroad. Silliman and Gibbs soon met, became friends, and spent time geologizing together around New England. In 1811, when he was considering a suitable place to display his mineral "cabinet," Gibbs made inquiries at institutions from Boston to Washington without quite finding what he was hoping for. Finally, he stopped to visit in New Haven, according to Silliman, and announced, "I will open

The Art of Misquotation

Over the years, many authors have happily repeated President Thomas Jefferson's supposed response to the report by Benjamin Silliman and James L. Kingsley about the Weston meteorite—that it was "easier to believe that two Yankee Professors could lie than to admit that stones could fall from heaven." But plenty of others have doubted the veracity of the quotation and devoted considerable effort to the attempt to track down its source.

The first published version of Jefferson's supposed remark appeared in the transcript of a speech by Silliman's son, Benjamin Jr., in 1874, a lifetime after the event. He called it "evidence of the limited knowledge of such subjects then prevailing in this country, even in the minds of the most cultivated people."[a]

That didn't sound right to Peabody Museum archivist Barbara Narendra, in part because Jefferson had himself been involved with a report on a meteorite in Baton Rouge just six years earlier. Dumas Malone, the eminent Jefferson biographer, agreed. He suggested that some careless writer had paraphrased and distorted Jefferson's cautious remarks about the Weston meteorite in a letter the president wrote in February 1808.[b] But for Narendra that didn't fit either. Paraphrasing was one thing. But where did that odd, flowery rhetorical form come from?

The oldest known approximation of Jefferson's actual response, termed in notably blunt language, eventually turned up in an 1826 memorial by former U.S. senator Samuel L. Mitchill of New York. Mitchill had received the earliest news of the Weston meteorite in a letter from an entirely different Connecticut duo who had investigated the scene just before Silliman and Kingsley. A fellow senator who had plans to dine with Jefferson the day the letter arrived asked if he could show it to the president. By hearsay, Mitchill reported that the letter had elicited "scornful indifference" from Jefferson, who said "he could answer it in five words . . . *It is all a lie.*"[c]

So where did Benjamin Silliman Jr. get his version of the quote? Having studied Benjamin Silliman Sr. for years, Narendra knew him almost as a friend, and suddenly one day it hit her: that flowery rhetorical phrasing was pure Silliman Sr.

Memory casts a fog over the past, and events that happened to other people sometimes unwittingly become our own. In any case, what harm would there have been in appropriating—and considerably embellishing—the mistaken remarks of such an eminent president? Silliman Sr. might not have done it in print. But as a tale told around the family dinner table and related to decades of students, why not? His devoted son's only error had been to believe it was true.

No. XV.

AN ACCOUNT OF THE METEOR,

Which burst over Weston in Connecticut, in December 1807, and of the falling of Stones on that occasion.

BY PROFESSORS SILLIMAN AND KINGSLEY.

WITH A CHEMICAL ANALYSIS OF THE STONES,

BY PROFESSOR SILLIMAN.*

ON the 14th of December, 1807, about half past 6 o'clock, A. M. a meteor was seen moving through the atmosphere, with very great velocity, and was heard to explode over the town of Weston, in Conne

* NOTE....The following account of the facts whic falling of stones from the atmosphere, was first p *stance*, in the Connecticut Herald, and, subseque papers, and in several literary and philosophic ed account, together with the details of the communicated to the Philosophical Soci been published in their transactions. to the Connecticut Academy, becau publishing any thing *immediatel* much alive on a subject which, el, that there was no room fo But, in consequence of t Connecticut, as the scene Academy have thought p papers, that they may b disclaiming at the same t nications.

This hefty stone made a sensation on plummeting down in western Connecticut, and again when Silliman Sr.'s chemical analyses indicated that it had an extraterrestrial origin.

it here in Yale College, if you will fit up rooms for its reception." Yale promptly did so, on the second floor of what is now Connecticut Hall.[13] And thus, among many other treasures, the 36.5-pound Weston meteorite came to Yale.

Gibbs provided one other critical boost not just to Yale but to American science at large. Late in 1817, he bumped into Silliman by chance one day aboard the steamboat *Fulton* on the ten-hour run between New York and New Haven. A mineralogist who had sporadically published a journal for that discipline was in failing health, and Gibbs urged Silliman to take up the challenge of producing a more broadly focused scientific journal. The following year, Silliman launched the *American Journal of Science*. It soon became the nation's premier scientific periodical, often referred to simply as "Silliman's Journal." Thomas Jefferson, nearing the end of life, would rank it among the "select reading for which alone I have time now left."[14]

The rapidly growing Yale mineral collection—in truth, still largely the Gibbs collection—meanwhile began to attract important visitors to New Haven. The

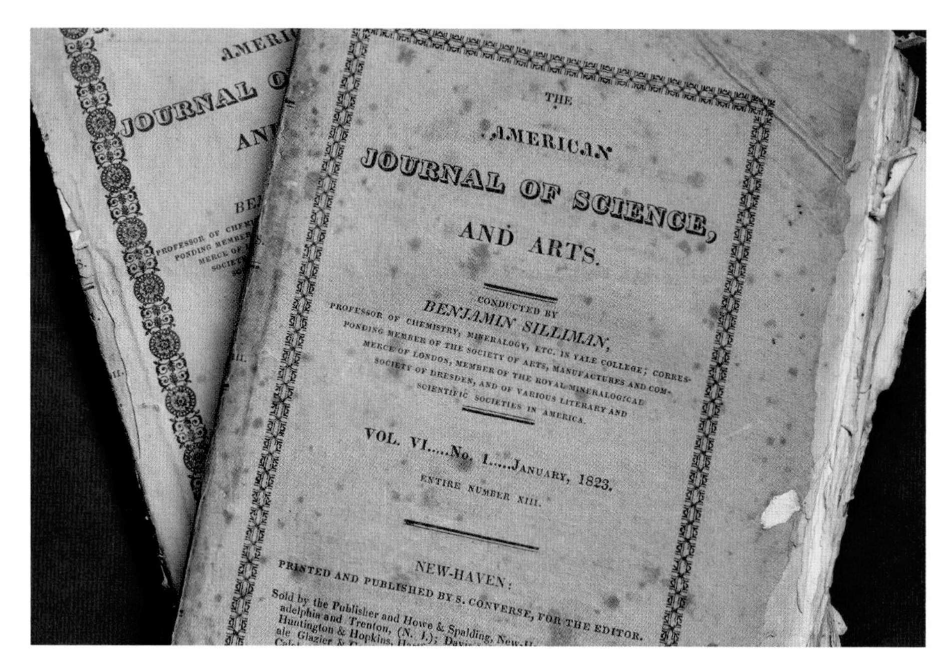

In 1818, Silliman Sr. founded the *American Journal of Science and Arts.* Once known simply as "Silliman's Journal," it is now the oldest continuously operating science journal in the United States and still published at Yale.

collection moved in 1820 to a space upstairs from the new college dining hall, a prominent building in the heart of the campus later known simply as the Cabinet Building. For Silliman and Yale, things seemed to be progressing smoothly. But in 1825, Gibbs suddenly announced that he needed to sell, his spending habits having caught up with him for the moment. The price Gibbs named for the mineral collection, $20,000, represented two-thirds of Yale's annual income. Silliman was soon out raising funds by pamphlet, public meeting, and door-to-door subscription in New Haven and New York. Yale's new president, Jeremiah Day, also knocked on doors, determined not to lose the collection, which had been, said Silliman, "for so long our pride and ornament."[15] In the end, they raised half the asking price, and Gibbs graciously accepted a note for the balance.

Silliman's Legacy

The collection became the basis around which a community of scholars—scientists, as they were just beginning to be known—began to gather at Yale. Silliman built on the prestige of the collection to help found a science school (later the Sheffield Scientific School), the first college art gallery (largely by arrangement

In 1802, in a candle box like this, Silliman Sr. carried molybdenite, chalcopyrite, hematite, and other mineral specimens to be properly identified in Philadelphia.

with his wife's uncle, the artist John Trumbull), and the medical school, helping nudge Yale well along the path from a college to a university. Among those who turned their attention to Yale as a result was an amateur mineralogist still at prep school named O. C. Marsh, who carefully noted in his journal a quotation from Silliman on the pursuit of acquisitions: "Never part with a good mineral until you have a better."[16]

Silliman's influence also radiated out from Yale. One of his students, Daniel Coit Gilman, would become the college librarian and an important figure in the rise of the Sheffield Scientific School. He would later take Silliman's training and move on to found Johns Hopkins University and the Carnegie Institution, both with a focus on scientific research. Another Silliman student, Amos Eaton, would help found what is now the Rensselaer Polytechnic Institute and pioneer the practice of applying the scientific method "to the common purposes of life."[17]

Silliman accomplished so much in part by establishing a family dynasty in science. Daniel Coit Gilman was distant kin: his brother had married a Silliman daughter. George J. Brush, a mineralogist and later director of the Sheffield Scientific School, had married Harriet Silliman Trumbull, a Silliman cousin. The faculty also included Silliman's son Benjamin Jr., a chemist who would play a key role in launching the age of petroleum, and Silliman's former student and son-in-law James Dwight Dana, who was both a zoologist and a renowned geologist. Dana's son and Silliman's grandson Edward S. Dana would also eventually join the faculty. What mattered, beyond the nepotism, was that Silliman had made it seem perfectly reasonable for Yale, the former Congregational seminary, to employ eight faculty members in science compared to just five in theology.[18]

The line dividing science and theology was, however, still practically nonexistent. On July 30, 1856, James Dwight Dana, who was both a deeply religious man and the greatest American geologist of the nineteenth century, gave a speech to Yale alumni lamenting those "who still look with distrustful eyes on science." They seem, he said, "to see a monster swelling up before them which they can not define, and hope may yet fade away as a dissolving mist." That specter was twofold: Geology had cast a shadow on the Genesis account of the Earth's history, and the idea of evolution, which was already in the air, had raised doubts about species as separate and divine creations. (Among other developments, a former student of Silliman's named Thomas Staughton Savage, who was a missionary and medical doctor, had recently brought home from Africa the bones of an unknown primate with a disturbing resemblance to humans. He named it "gorilla.") But Dana was deeply committed to a biblical view of creation, and he assured his listeners that geology in fact provided evidence "that God's hand, omnipotent and

The Rise of the Age of Oil

College professors were notoriously underpaid in the nineteenth century and often had to patch together a living by hiring themselves out as experts. Benjamin Silliman Jr. did so with more gusto than most, advising mines, quarries, and public water systems, helping to found the New Haven Gas Light Company, and serving as president of the Connecticut Steam Heating Company, among many other endeavors.

His tendency to optimism on behalf of clients and his frequent carelessness about the underlying details would eventually cause critics to mutter that he was cashing in on the Silliman and Yale names. James Dwight Dana remarked privately that his brother-in-law, neighbor, and coeditor of the *American Journal of Science* was "on the constant go in behalf of one thing or another, and alas for Science."[a] But one of the reports Silliman wrote as a consultant was spectacularly right in a way that changed the history of the nation and the planet.

It started with a patent medicine, Kier's Genuine Petroleum or Rock Oil, said to cure everything from diarrhea and deafness to pimples.[b] It was also flammable. To enterprising investor George Bissell, this suggested that the "rock oil" seeping from the ground in northwestern Pennsylvania could serve not just to make patent medicine but also to illuminate private homes as a replacement for whale oil and a competitor to kerosene and "coal oil." He formed an alliance with a New Haven banker, James M. Townsend, and they hired Silliman to provide an analysis.

A sample of the brownish oil, like thin molasses, arrived at Yale late in 1854, and Silliman set him himself diligently to the task. Fractional distillation involved heating a sample in one vessel of an alembic, then collecting the resulting distillate in a connected vessel. Different temperatures yielded different chemical components in the oil, some arguably suitable as lubricants, others for illumination. "I am very much interested in this research," Silliman confided to Bissell that December, "and think I can promise you that the results will meet your expectations of the value of this material for many most useful purposes."[c]

A series of delays ensued, and an explosion in the lab required the investors to purchase new equipment. The different analyses Silliman made also took much longer and cost more than the anxious investors had hoped. Also anxious, Silliman withheld his report that April until payment arrived in his account. The cost to Bissell, Townsend, and company totaled $600—roughly $16,200 in today's money. But they got what

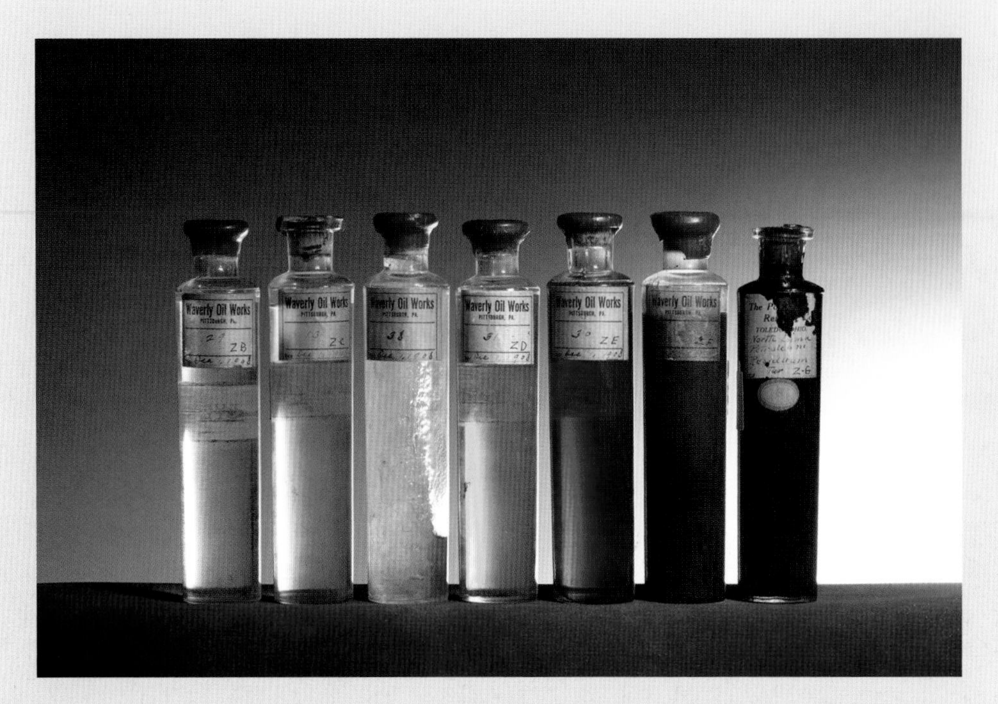

Silliman Jr. spent months in the mid-1850s distilling and analyzing crude oil from northwestern Pennsylvania. He produced distillate samples like these, helping to launch the age of oil.

they needed. Without ever having visited the site in northwestern Pennsylvania, Silliman found that rock oil "exists in great abundance upon your property" and would prove "unfailing in its yield from year to year." More authoritatively, he promised that a relatively simple refining process could produce "the most highly illuminating of all the carbon gases."[d]

This report attracted the necessary capital to begin work, and the investors soon reorganized as the Pennsylvania Rock Oil Company, with Benjamin Silliman Jr. as president. They borrowed their proposed drilling technique directly from the image of drilling derricks on the label of Kier's rock oil patent medicine. To build the derricks and manage work on the site in Titusville, Pennsylvania, Townsend found a likely candidate in a fellow resident of the Tontine Hotel, overlooking the New Haven Green. Edwin L. Drake was a railroad conductor and odd-jobs man with no visible qualifications. But he possessed an engaging manner and what would prove an indomitable will to accomplish the task before him.

Drake became Colonel Drake for the purpose of impressing the people of Titusville, and he spent much of 1858 and 1859 hard at work there. (He also replaced Silliman as president in another reorganization, as the Seneca Oil Company.) The available capital ran out before any oil appeared, and for a time Townsend kept Drake going out of his own pocket before reluctantly sending a letter in August 1859 telling him to shut down the operation. But before the letter reached Titusville, Drake's drillers hit a vein and were quickly producing fifteen to twenty-five barrels of oil a day. Within a year, this new fuel was being marketed as "the light of the age," almost a match for "the clear, strong, brilliant light of day." For better or worse, Silliman's report had launched the age of oil.[e]

bearing a profusion of bounties, has again and again been outstretched over the earth; that no senseless development principle evolved the beasts of the field out of monads,"—that is, unicellular organisms—"and men out of monkeys, but that all can alike claim parentage in the Infinite Author."[19]

Having dismissed the evolutionary bugaboo, Dana went on to argue for the expansion of scientific study on the Yale campus, with new laboratories, lecture halls, and above all a museum, "a spacious one." This museum, Dana said, "should lecture to the eye, and thoroughly in all the sections represented, so that no one could walk through the halls without profit. It should be a place where the public passing in and out, should gather something of the spirit, and much of the knowledge, of the institution."[20]

Then rising to his conclusion, he called on the alumni to help build "the first university in the leading nation of the globe. . . . Why not have here, in this land of genial influences, beneath these noble elms . . . why not have here, THE AMERICAN UNIVERSITY,—where nature's laws shall be taught in all their fullness, and intellectual culture reach its highest limit!"[21]

By coincidence—or perhaps he would say "miraculously"—the means of making the museum Dana imagined a reality entered the college community the day after he delivered that speech. On July 31, 1856, aspiring freshman O. C. Marsh took the exam for admission to Yale. By good fortune for all, he aced it.[22]

CHAPTER 3

The Education of O. C. Marsh

I have spent enough time shooting ducks to fit myself for college.

O. C. MARSH, 1852

Unlike many of the students who traveled with him in 1870 and on subsequent expeditions, O. C. Marsh had grown up in modest circumstances. He was a New York State farm boy from Lockport, north of Buffalo. His childhood had been extended—he was twenty-four before he started college—and seemingly unhappy. His father, Caleb, was undependable, financially and otherwise. His mother, Mary Peabody Marsh, died when he was just three. Othniel and his older sister Mary then went off to live with relatives north of Boston, rejoining their father only after he remarried a few years later. At that point, the two siblings became stepchildren in a family that soon grew by six more children. With so many mouths to feed, Caleb pushed his eldest son to work the fields, and he punished him harshly for his lapses.

Othy, as he was then known, had "a roving disposition" and slipped away at every opportunity to fish, hunt, and geologize. Excavations to widen the Erie Canal were under way just a mile from the family farm. The heaped-up rock—Lockport dolomite and Rochester shale—proved to be loaded with various minerals as well as with fossil corals, bryozoans, and trilobites. A neighbor, Colonel Ezekiel Jewett, who had made a reputation as a mineralogist and paleontologist, gave Marsh his first lessons in what the rocks and the landscape might reveal. Jewett was also a collector of Native American ethnological material, another interest Marsh was to pursue in later life.[1]

That informal schooling—and his mother's family, the Peabodys—would eventually rescue Marsh from the hardscrabble life of a farmer or carpenter. His uncle George Peabody had given a dowry to his favorite sister Mary Peabody when she married Caleb Marsh. With his characteristic prudence, Peabody had stipulated that it pass to her children in the event of her death. By the time O. C. Marsh came of age in 1851, his share amounted to $1,200. At that point, the goldfields of California or some other distraction might easily have beckoned. But guided by an aunt, probably Judith Peabody Russell, Marsh went east instead to enroll in the Phillips Academy in Andover, Massachusetts. "I think he intends to remain there through the spring and then work at some mechanical trade," his sister Mary, by then married and also living in Massachusetts, wrote home to their father.[2]

O. C. Marsh in about 1860, during his student days at Yale College.

Marsh seems to have taken his studies lightly that first year, with a special focus on backgammon. But then Mary died suddenly in her twenties, as their mother had, and the loss shook O. C. Marsh badly. By all the evidence, it was his last deep emotional bond with another human being, at least of the positive variety. Soon after, looking out over the Massachusetts countryside from Dracut Heights in Lowell, Marsh determined that he would henceforth become serious about the course of his life—that is, about his work. "I have spent enough time shooting ducks to fit myself for college," he remarked.[3]

Marsh went on to excel as a scholar, eventually becoming valedictorian of his class at Andover. He was also captain of the football team and a volunteer fireman. (His journal includes a vivid account of fighting a barn fire.) In summers, he went on mineralogy expeditions to Nova Scotia. He seems to have been adept at forging useful connections among his schoolmates, as he would be later in the worlds of science and government. A classmate recalled being impressed by Marsh's "superiority in managing practical affairs" and by his foresight and shrewdness, "with a touch of cunning in it," in securing the presidency of the debating society.[4] The most useful connection for young Marsh was of course his own uncle, George Peabody, who was already supporting him at Andover. Under his aunt Judith's watchful eye, Marsh courted him eagerly.

Uncle George

Though he is remembered hardly at all today other than as a disembodied name on his many philanthropic bequests, George Peabody was among the most admired figures of the mid-nineteenth century on both sides of the Atlantic. Born in 1795, the third of eight children, Peabody grew up poor in Danvers, Massachusetts, north of Boston. He received little proper schooling, a source of lifelong regret, instead going to work in the village store at the age of eleven. His father died when he was just sixteen, and George began to support his mother and younger siblings, erasing the family debt while he was still in his early twenties. From there he went on to build a massive fortune, first as a retailer in Baltimore, then trading copper, silk, tea, and other commodities internationally, and later as a merchant banker in London. The firm he founded there thrived by attracting European capital for American railroads, canals, and other enterprises. It would become, after he retired and withdrew his name, J. S. Morgan & Co., also known as "the House of Morgan," antecedent of JP Morgan Chase, Morgan Stanley, and other present-day financial powers. Finally, in the last decades of his life, he gave away much of his fortune. In the process, he became the father of modern philanthropy and the model for donors from Andrew Carnegie and John D. Rockefeller to Warren Buffett and Bill Gates.

Peabody is, however, a difficult figure to come to grips with today, partly because he was the subject in his lifetime of such ardent hagiography. One writer then praised him as "brave and noble in thought and action, lofty in purpose, and prompt whenever the call of duty came." Preachers sermonized his generosity from the altar, and even the novelist Victor Hugo eulogized him as one of those men in whose face "we can see the smile of God."[5]

In London, Peabody made himself the agent of all things American, regularly hosting lavish Independence Day parties and serving as a genial guide to visiting Americans. When Congress balked in 1851 at spending $15,000 to support the U.S. display at the Great Exhibition in London's Crystal Palace, Peabody advanced the money instead. The resulting display, featuring Colt's revolvers, Cyrus McCormick's reaping machine, and other examples of American ingenuity, was a sensation—and no doubt also good for Peabody's business.

Despite his small army of grateful admirers, Peabody seems to have had little in the way of a private life. A brief engagement to a much younger American woman in 1838, when he was forty-three, ended unhappily. A persistent rumor held that he kept a mistress in Brighton and had a daughter with her. Two supposed grandsons sometimes showed up at the company's office decades later, according to a statement left in 1940 by J. P. Morgan Jr., "and both of them had the old man's nose to a dot." But the official position was that Peabody did not waste "his energies in . . . the gratification of the lower nature," nor was there any hint of "a frittering-away of his powers in unmanly amusements and senseless frivolity. He was no mere pleasure seeker." Hence perhaps one bon vivant's remark that Peabody, known to American visitors as "Uncle George," was "one of the dullest men in the world: he had positively no gift except that of making money."[6]

Peabody worked relentlessly and lived modestly, preferring furnished rooms at a hotel to the sort of mansion he could well afford. The *Boston Post* described him as "eminently a peculiar man" who sometimes carried his "inveterate dislike" of display "to a ridiculous length." At one point, the paper reported, Peabody stood patiently in "a drenching storm because he preferred a horse-car to a hackney-coach," that is, a bus to a taxi. Historian Ron Chernow characterized him flatly as "Scrooge," sitting atop a fortune estimated at $20 million.[7]

Peabody's extensive philanthropy did not, however, spring from any last-minute epiphany induced by ghostly visions of Jacob Marley's chains. On the contrary, he claimed to have planned "from the earliest of my manhood" to give away the bulk of his fortune. The emphasis on education also seems to have come early, motivated by his own painful lack of it. In an 1831 letter to a nephew who was seeking financial help for college, Peabody wrote that he was "*well qualified* to estimate" the value of education "by the *disadvantages*" he faced without it, both at work and in society. He added that he would "gladly pay *twenty times* the expence attending a good education could I possess it." But it was too late for him. Instead, he paid to educate members of his family, beginning as a young man with his siblings, including his sister Mary, who regarded him as "more than a father," and then with various nephews and nieces. In 1856, it came time for Mary's son O. C. Marsh to move on to college. By letter, Marsh thanked Peabody for supporting him at Andover and, he hoped, in college. He assured him he would "never have occasion to regret the kindness you have shown to me for my dear mother's sake." The Peabodys had customarily been Harvard men, but Marsh was set on Yale because of the strength of its science department. Peabody did not quibble. Education was the thing.[8]

When Marsh described his first fossil species, *Eosaurus acadianus,* in 1862, he showed his budding competitiveness, flatly contradicting the opinion of eminent Harvard zoologist Louis Agassiz.

It was Marsh's first new species and an opportunity to boast to his uncle that his work "has already attracted considerable attention among scientific men."[7] In that same letter, Marsh asked to complete his education in Germany, and Peabody not only approved but this time sent the money directly rather than through his sister Judith. (With the Civil War under way, Marsh was offered an army commission as a major but was disqualified because of defective vision.)

That summer Marsh's hometown paper reported that Marsh was expecting a professorship at Yale and that Peabody was not only paying his expenses but intended to make him his heir. Marsh fired off a furious letter to his father, the apparent source, berating him for remarks that were "injudicious . . . entirely uncalled for," and likely to undermine the very things they celebrated.[8] Luckily, neither Peabody nor Marsh's professors at Yale seem to have read the *Lockport Journal and Courier.*

In November, Marsh was in London meeting Peabody for the second time. They hit it off. When Marsh hinted that he had a request to make, Peabody smiled and said, "For money, I suppose?" Marsh replied, "Not a penny for myself, but I wish you to endow Science at Yale College." According to an author visiting then in London who said he had heard an account of the meeting from Marsh, "that reached Peabody's heart and his purse."[9]

Marsh, now thirty-one, was turning out in many ways to resemble his famous uncle, and as they began to know each other personally, they seemed to find camaraderie in their kinship. Both felt the same powerful urge to accumulate (money for Peabody, fossils and minerals for Marsh), and both were shrewd and sometimes ruthless in pursuit of their goals. Both intended, paradoxically, to give away everything in the end. Neither felt much need for close human contact or for marriage, though both enjoyed having large numbers of useful acquaintances. Each found comfort instead in the tantalizing discipline of work.

Marsh clearly held his uncle in awe and aimed to emulate him. "Before I retire I should like to do for Science as much as you have done for your fellowmen," he wrote, "and if my health continues I shall endeavor to do so." The admiration was mutual. Education was always Peabody's Rosebud. The pain of having lost the opportunity for schooling in his own youth was palpable in the words with which he opened an 1831 letter on the topic: "Deprived as I was . . ." Now, though, through his favorite sister's child, Peabody found a direct and personal connection with scholarship, with exploration, with discovery. He took a keen interest in his nephew's scientific work, and Marsh was no less keenly attentive to Peabody's "future plans and donations."[10]

"My Dear Mr. Marsh"

"I will tell you confidentially that Harvard will have her usual good fortune," Marsh wrote to Benjamin Silliman Jr. that November. "So many of our family have been educated at Harvard that [Peabody] naturally felt a greater interest in that institution than in Yale, of which I am the only representative." Peabody had already talked with Harvard officials about founding a school of design, or art, there. But Marsh entertained hopes for Yale, noting that his uncle "manifested so much interest in my scientific studies that I thought it not unlikely that he would be more inclined to that department" than to art. Without having made any specific proposal to his uncle, he asked Silliman how a $100,000 donation might "be best employed for the benefit of science at Yale." Silliman, as it happened, had just such a plan in his back pocket, and he suggested that a new science building "bear the name of the founder in such terms as may be most permanent and expressive—e.g., 'Peabody Museums' or 'Peabody Hall.'" He thought such a project would cost upwards of $150,000.[11]

Marsh warned against making a "direct application" to Peabody. The best approach with his uncle was simply to plant the seed of an idea and leave it to germinate of its own power. Marsh turned to his studies in Germany, and nothing more happened until the following spring. Then Peabody, taking the cure at the spa in Wiesbaden, summoned his nephew for a visit. Soon after, Marsh wrote to the elder Silliman, "I take great pleasure in announcing to you that Mr. George Peabody has decided to extend his generosity to Yale College, and will leave a legacy of one hundred thousand dollars to promote the interests of Natural Science in that Institution." He held out the possibility that the donation might ultimately increase and come in Peabody's lifetime rather than as a legacy.[12]

James Dwight Dana responded ecstatically: "My Dear Mr. Marsh. Your good words and Mr. Peabody's generous deeds have filled us with rejoicings. . . . I can almost see the grand structure standing in its place." It should be big enough, he imagined, to accommodate an expanding student body, "and it ought to be made to last at least 1000 years. . . . It will be a great day for Yale when the building is completed!" As an aside, he mentioned that he thought there would "be no difficulty as to your appointment" to the Yale faculty.[13]

The younger Silliman was less lyrical and more pragmatic. He advised Marsh to "devote yourself with zeal and your well known perseverance to the subject of Paleontology" instead of his intended focus on mineralogy. "Paleozoic rocks abound in the U.S. and demand by far more study than they have received." Not incidentally, paleontology was the only field of scientific study for which Yale had a faculty position available. In *his* aside, Silliman suggested that Marsh should persuade Peabody to "endow your Professorship" and—why not go for broke?—designate a fund "to sustain all the departments contemplated in our plan, there being no other means to sustain them and nothing more useless than costly establishments without foundations."[14]

Meanwhile, Marsh wrote to Peabody that his "munificent donation to Yale" had helped him secure the promise of a faculty position at a rank corresponding "to that held by the great Agassiz at Harvard" and by the elder Silliman at Yale. He defined paleontology for his uncle ("the science of fossil remains") and requested $3,000 to $4,000 for the purchase of a library and a cabinet, or specimen collection, suitable to the post, noting that these essential tools were to a science professor "exactly what capital is to a man in business." Peabody wrote back approving $3,500.[15]

Marsh was evidently content with the shift to paleontology. "So much for mineralogy," he wrote, with only mild regret. Then he turned to his studies in earnest, concerned now not just with fulfilling his own and his uncle's high expectations but with preparing himself "to fill the position with honor to the University." He expressed his worries more plainly two years later. While visiting London after

Marsh collected these invertebrate fossil specimens in Germany (*clockwise from upper left*): a ghost-shrimp, preserved excrement possibly from a cephalopod, a shrimp's forelimbs, a crinoid, and the hooks on the tentacles of a squid.

completing his studies in Germany, he found himself unable to enjoy the opera and theater tickets his uncle had left for him. Instead, he spent his time at work in the British Museum, "fearing that I might not know enough to be Professor at Yale."[16]

Marsh would be returning to a Yale with grander ambitions than the one he had left just three years earlier. The Sheffield Scientific School was expanding its faculty and becoming a greater presence on campus, partly because of the continuing support of its namesake, New Haven railroad magnate Joseph E. Sheffield (whose son-in-law John Addison Porter taught chemistry there). Yale awarded the first Ph.D. in North America to a Sheffield student in 1861. It was also one of the first schools to take advantage of the Morrill Act of 1862, a national land-grant program promoting technical and scientific education. With the added promise of the Peabody gift, an exultant George J. Brush, who had been one of Marsh's professors, wrote to him in Germany, "You can form no idea how every one seems to have waked up to the fact that Yale is to be a great university." Dana agreed: "The time of her renaissance has come."[17]

Becoming Darwinian

Improbably, the most critical intellectual development of this period went almost entirely unmentioned. Charles Darwin had published his *On the Origin of Species* in November 1859, and it shattered the biblical interpretation of the Earth's history as if with a geological hammer, bringing the ghostly specter of evolution to glorious life. It is unclear when Marsh became a Darwinian, but the transition might logically have begun in Germany, when he was no longer merely a protégé of Dana and Benjamin Silliman Jr. A translation of Darwin's book appeared there just months after the English publication, and most German biologists accepted the idea of evolution by natural selection enthusiastically. But many of Marsh's own professors in Germany were adamantly in the rear guard. One of them, the comparative anatomist and microscopist Christian Gottfried Ehrenberg at Berlin University, regarded *On the Origin of Species* as a "completely mad book."[18]

The copy of *On the Origin of Species* in Marsh's own library is an 1866 edition. It seems likely, though, that his Darwinism developed earlier than that, and that he drew it from the source. On his first London visit in 1862, Marsh had characteristically sought out the right people, among them Charles Lyell, the leading geologist of the day. Both had worked on the Joggins Fossil Cliffs in Nova Scotia, so it was natural to make the connection, and either Dana or Peabody's name would have smoothed the way. (That December, Lyell passed Marsh's *Eosaurus* paper along to be read before the Geological Society of London.) Lyell was Darwin's close friend and ally.

At some point, Marsh also became acquainted with Thomas Henry Huxley, "Darwin's Bulldog." Later, he was a guest at the legendary X Club, at which Huxley and other leading thinkers met to eat and argue, united by their "devotion to science, pure and free, untrammelled by religious dogmas." Later still, Marsh visited Darwin himself at Down, his house outside London. And yet Marsh left no record of what they discussed or what he thought of the revolutionary theory.[19]

Marsh's silence on the subject was no doubt strategic in the first half of the 1860s, when he was negotiating with the creationist holdouts at Yale to establish himself and his museum in their midst. But even when Marsh declared his allegiance to evolutionary thinking unmistakably, it was, oddly, as an argument for not talking about it out loud. "But I am sure I need offer here no argument for evolution," he told the American Association for the Advancement of Science in 1877. Before the same group in 1879, he remarked that most scientists thought it "a waste of time" to discuss the truth of evolution, adding somewhat optimistically, "The battle on this point has been fought, and won."[20]

The Importance of the Natural Sciences

A hint survives, nonetheless, that Marsh was frank with his uncle about his changing beliefs. Even as he was working to ensure that Peabody would properly "endow Science at Yale," Marsh was also maneuvering to establish the exact terms of Harvard's "usual good fortune." In place of the original plan to found an art school there, Peabody had shifted under Marsh's influence toward science in the form of a possible gift for the Harvard observatory. But Marsh and his Harvard cousin, George Peabody Russell, proposed instead that he found a museum of American archaeology and anthropology.

Marsh visited Harvard early in 1866 to sound out this plan, but not with university administrators nor with Harvard's most eminent scientist, Louis Agassiz, a leading opponent of Darwinian theory. Instead, he went to botanist Asa Gray, who was at that time the nation's leading Darwinian. Gray welcomed Marsh's proposal. So did James Walker, the former Harvard president. With a faint shrug, Walker wrote, "I have always been of the opinion that when a generous man like Mr. Peabody proposes a gift, we should accept it on his plan, and not on ours."[21]

On October 8, 1866, during a return visit to the United States, Peabody agreed to donate $150,000 to create Harvard's Peabody Museum of Archaeology and Ethnology. The founding letter, with its call for the museum to be attentive to any evidence in America of "human remains or implements of an earlier geological period than the present," might have been a simple call to pursue the new science of anthropology.[22] But it also suggests that Peabody was himself attentive to the evolutionary debate over human origins.

Two weeks later in New Haven, on October 22, 1866, Peabody declared that he had become convinced "of the importance of the natural sciences." He was therefore committing his financial support, also for $150,000, to "the foundation and maintenance of a MUSEUM OF NATURAL HISTORY" at Yale, emphasizing zoology, geology, and mineralogy.[23]

O. C. Marsh was now thoroughly established both as the professor of vertebrate paleontology at Yale and as a curator and trustee of the Peabody Museum. In a note sent from Germany the year before, he had wondered pointedly whether the plan being developed by Dana and Silliman was "large enough for the requirements of the future." What he had seen of those plans "would certainly not be large enough for the present Berlin collections." Never mind that Berlin was the capital of a major European nation and home to about 650,000 people, while New Haven had just 45,000 residents. Marsh was unabashedly ambitious, asking, "Should not [the New Haven collections] soon be as extensive?"[24] Bowing to that expansive vision, the trustees of the new museum decided to delay construction while their available funds gathered interest.

Peabody, meanwhile, left the details of the new museum to be worked out by Marsh, Dana, and their colleagues, turning away from science to attend to his other philanthropic causes. He would live another three years, still traveling but increasingly limited by various ailments. He died in November 1869, and the mourning on both sides of the Atlantic was evidence of just how new and different his philanthropy seemed to the world then. In his lifetime, Queen Victoria had offered him a title, which he had declined as ill suited to an American, and smaller honors, including a portrait of herself and a note praising him for generosity "wholly without parallel," which he had proudly accepted.[25]

Now the queen gave permission for him to be buried among the poets and kings in Westminster Abbey pending the repatriation of his body to the United States. The elite of the British Empire, no doubt including some who had regarded "Uncle George" with snobbish disdain in life, crowded into Westminster for the funeral, and the *New York Times* reported that many of them wept. Outside in the cold, "the gaunt, famished London poor were gathered in thousands to testify their respect for the foreigner who has done more than any Englishman for their class."[26] The following January, Britain assigned its newest and largest warship, the HMS *Monarch,* to carry Peabody's coffin across the Atlantic. Not to be outdone, President Ulysses S. Grant directed a U.S. warship to serve as an escort, and at the end of January 1870, an armada went out from Maine to welcome George Peabody home.

George Peabody's name appears on museums and philanthropic institutions, but the man himself, shown in an 1867 portrait, is largely forgotten.

Aftermath

One day the following August, in the dismal frontier village of North Platte, Nebraska, someone looked up to notice a cloud of dust approaching across the plains. As a large party of men riding Indian ponies galloped closer, "the wildest consternation" seized the villagers. They ran for their weapons, terrified at what appeared to be an attacking band of Sioux. In fact, it was O. C. Marsh, his Yale paleontological party, and their military escorts, endowed by two weeks of fieldwork with a "wild and savage appearance" and a momentary impulse to mischief. The temporary confusion resolved itself without gunfire.[27]

The first of the four explorations Marsh was planning for that year had passed successfully. The Sioux had in fact followed this puzzling party from a distance, setting fires that forced the men of the expedition to march across burned prairie, "studded with roasted cactus and dead grasshoppers" and without adequate grass to feed the horses. The temperature had at times risen to 110 degrees in the shade. ("Why *did* God Almighty make such a country?" one soldier lamented, and another replied, "God Almighty made the country good enough, but it's this deuced geology the professor talks about that spoiled it all!") They had persisted, and after a time, even the soldiers and the Pawnee scouts had joined in the digging. (The Pawnee had supposedly dubbed Marsh "Bone-Medicine Man" and also "Heap Whoa Man," the latter for the peculiar, somewhat hesitant, interjections he directed at his horse, probably as he was studying the ground for fossils.) They had discovered two American rhino species and six primitive horses, and shipped thirty-six boxes of fossil discoveries to the Peabody, including an estimated one hundred extinct vertebrate species previously unknown to science. It was the beginning of what would soon become one of the great collecting sprees in the history of science.[28]

CHAPTER 5

Rock Render

To change is always seeming fickleness. But not to change with the advance of science is worse; it is persistence in error.

JAMES DWIGHT DANA, *System of Mineralogy*

Anyone living in New Haven in the second half of the nineteenth century might, with luck, have stumbled upon the Saturday afternoon spectacle of a professor heading out from the Yale campus at a clip, hotly pursued by fifty or more students armed with geological hammers. The students were soon wielding their weapons "with great vigor on the boulders which strew the New Haven plain" and presenting their prizes to their teacher, James Dwight Dana, for his patient elucidation. Dana was, as one of those students recalled him, a slightly built figure of medium height with "keen blue eyes" and "a thin, sensitive face with finely carved features" under a thick mass of hair. His standard expression was "a kindly smile, touched with a trace of humor." Dana was also quick: he had too much to show his students in a single afternoon derived from his ceaseless explorations of the area's trap ridges, kettle holes, and boulder trains left behind by the retreating North American ice sheet. Racing from one point of interest to the next, Dana "would break into a sharp trot, which the best sprinters present did not care to outdo for very long." By the time the group straggled home at dusk, only a dozen or so remained of the fifty or more "who had so bravely started."[1]

Dana was no ordinary professor, entirely apart from his "unconquerable energy." He was, among many other things, the Linnaeus of the geological world. The Swedish botanist Carolus Linnaeus had devised the beginnings of a system for classifying and naming living things by genus and species, and Dana's system did much the same for minerals. Remarkably, Dana later categorically "rejected" his

James Dwight Dana made astute observations of the natural world but struggled to reconcile them with his deep faith in the Bible's account of creation.

own system "as false to nature" and built a better one based on the chemical properties of different minerals. "To change is always seeming fickleness," he wrote. "But not to change with the advance of science is worse; it is persistence in error."[2]

Dana published his 580-page *System of Mineralogy* in 1837, when he was a twenty-four-year-old assistant in Benjamin Silliman Sr.'s laboratory—and then relentlessly revised and improved it. Even now, almost 170 years later, Dana's classification scheme is standard in the geological world, and colleges still teach his *Manual of Mineralogy* and his *Manual of Geology* as essential texts in the field.

This would have been plenty for any scholar to have accomplished in one lifetime. But Dana also made himself the world's leading expert on corals, crustaceans, and volcanoes, publishing massive volumes on each, largely with his own illustrations. Along the way, he sailed round the world; lived and worked for a time among cannibals; survived a shipwreck and other perils of a nineteenth-century expedition; described seven hundred new species; married Henrietta Silliman and raised four children with her; served for many years as chief editor of Silliman's *Journal* (and indexed it); kept up a vast correspondence; helped establish the Sheffield Scientific School and in the process reorganize Yale University; led planning of the Peabody Museum; played guitar and flute; and composed a fifty-six-page songbook including a complete Nativity.

Dana characterized himself as a gatherer of neglected facts that would sooner or later find their place in science. "It is sufficient for me to have studied them out, and given them to the winds," he wrote to a friend in 1851. Dana also drew his facts together to form audacious theories about the basic laws of nature. As his biographer and Yale contemporary Daniel Coit Gilman, later president of Johns Hopkins University, put it, "The study of a rock, of a crystal, of a crustacean . . . or of a continent led upward and outward to the mysteries of the universe, to the origin, the order, and the purpose of the world."[3]

If that sounds reminiscent of Charles Darwin, it should. Though Darwin and Dana lived on opposite sides of the Atlantic, their lives ran on similar lines and sometimes crossed paths. Both, for instance, published lengthy books on different branches of crustacean taxonomy at about the same time. (Dana's came soon after his weighty volumes on corals and on geology, leaving Darwin dumbfounded: "Why, if you had done nothing else whatever, it would have been a *magnum opus* for life! . . . I am really lost in astonishment at what you have done in mere labour. And then . . . so much originality in all your works! It frightens me to think of it.")[4] Both were interested in the same subjects — the formation of coral reefs and atolls, the forces shaping the rise of continents and the upthrust of mountain ranges, and of course the extraordinary abundance and character of species. Both made great voyages and perhaps as a consequence suffered chronic disabling maladies of mysterious origin, and both were also deeply engaged, though on opposite sides, in the question of evolution.

The U.S. Ex. Ex.

In August 1838, a few years after Darwin returned to England aboard the HMS *Beagle,* Dana went to sea, age twenty-five, on the USS *Peacock,* one of six vessels making up the U.S. Exploring Expedition, known popularly as the U.S. Ex. Ex. or the Wilkes Expedition for its commander, navy lieutenant Charles Wilkes. It was the nation's first great global scientific and geographic expedition on what would ultimately be a four-year circumnavigation of the globe. The *Peacock* was a 650-ton, three-masted sloop of war, and Dana, serving as the expedition's mineralogist, geologist, and de facto naturalist, was lucky enough to have a private stateroom, six by seven and a half feet and crammed with scientific gear. He was eager to work. With the ship caught in doldrums en route to Brazil, he collected and described ninety-five new species of marine crustaceans, mostly copepods, remarking, "The ocean contains yet more new things than either philosophy or science has hitherto dreamed of."[5]

The ocean, of course, also offered innumerable opportunities for disaster, and the most dangerous of them for Dana (other than the time he sank outright; see the illustration below) occurred just a few months into the expedition. He had transferred to the supply ship *Relief* in March 1839 when it was assigned to investigate harbors in the Magellan Straits at the tip of South America. Dana thought it would give him a chance to study the abundance of islands there. But as *Relief* headed toward the western entrance to the straits, foul weather sent the wind howling through the rigging "with almost deafening violence." Sleet and haze blinded the sailors, but they "dashed on, plunging through the waves or staggering over them, and occasionally enveloped by their foaming tops."[6]

Suddenly, in midafternoon, a voice cried out, "Breakers under the bows!" Dead ahead "rude towers of naked rock" rose two hundred feet in height. "The heavy surges of the southern ocean rolled in against the rocks with frightful roar and tumult," Dana wrote, "and now and then dashed the spray over their summits," sending "white, thready torrents down the black rocks." The captain, calmly giving orders amid the chaos, brought the ship about but again found heavy breakers and hidden rocks just ahead. (The area was so stippled with them that some wit had dubbed it "The Milky Way.") Unable to find a passage through to open sea, the ship came about again and slipped past the Tower rocks to take up a hazardous anchorage on the leeward side of nearby Noir Island, an inauspicious name.[7]

The gale raged on that night, "with occasional squalls of extreme violence," Dana wrote. The ship "reared and plunged with each passing wave," raising the fear "that every lurch would snap the cables or drag our anchors," leaving the ship to splinter on the nearby rocks. When the gale dwindled next morning, Dana, the incorrigible naturalist, talked "of a ramble on shore as soon as the sea should go down." Instead, the tempest came roaring back in the afternoon. Finally, at dusk, the anchors gave way, and their cables "rumbled like distant thunder upon the rocky bottom."[8]

Dana now considered a different set of options in the event of sinking: freeze to death in the frigid waters, be smashed to bloody bits among the rocks, or somehow reach shore only to starve to death on the barren island. "His the happiest lot who was soonest dead," he concluded, a prospect made less terrible for him because of his deep religious faith. The dragging chains "sounded a dirge," and then the captain ordered the anchor lines cut and sent his crew scrambling into the rigging to loose the sails. For a few dreadful moments, everything hung in the balance. Then the ship slipped around the southern cape of Noir Island and, as the weather miraculously cleared, the *Relief* broke for the open sea. They had not perished, Dana thought, only because "a merciful God had planned it otherwise."[9]

In 1841, Dana's ship the USS *Peacock* sank at the mouth of the Columbia River, taking down precious specimens. Abandoning ship at night in rough seas, all hands made it ashore and continued on foot to San Francisco.

The Coral Forests

After stopping in Chile and Peru, pursuing mountain expeditions in both countries, Dana, back aboard the *Peacock,* sailed west to explore the Pacific islands. He spent much of his time gathering corals, often "by wading over the reefs at low tide, with one or more buckets at hand to receive the gathered clumps." Where the corals were too deep, he hired divers and went "floating slowly along in a canoe," studying the reef through the clear water and pointing out the specimens he wanted.[10]

To Dana, these "groves of the ocean" were as varied and beautiful as the rain forests of Brazil. Much as falling leaves build up the soil in the forest, coral debris broken up by wave action drifted down to form a reef bed on the ocean floor around islands. But it didn't decompose like leaves. Instead, this "enduring rock-material" compacted and became a seemingly permanent base on which other corals grew. It was the beginning of a promising new theory: barrier reefs and ring-shaped coral atolls had originated and built up on the fringes of volcanic islands, some of which had later subsided. Dana was no doubt dismayed on arriving in Australia in November 1839 to read a brief notice that Darwin had just proposed the identical theory. Dana generously remarked that Darwin's work "threw a flood of light over the subject, and called forth feelings of peculiar satisfaction, and of gratefulness to Mr. Darwin."[11]

But as a field researcher, Darwin was no Dana. Moreover, the itinerary of the *Beagle* often obliged him to rely on guesswork in writing up his 1842 monograph *The Structure and Distribution of Coral Reefs.* Dana had far greater opportunity to explore the coral reefs, and of course he did so with his usual boundless energy, often at considerable risk. "We have been threading our way for the past month among the reefs and shoals of the Feejee Islands, sometimes aground" and often close to being aground, he wrote to his friend the botanist Asa Gray in mid-1840. "We are now so accustomed to thumps against the reefs that they seldom interrupt me in my studies or investigations below." Ashore in Fiji, he worked amid islanders he described as "a cruel, treacherous race of cannibals," though "kindly disposed towards us" and with a fortunate preference for roasting and eating only one another ("a white man, they say, tastes bitter"). His constant collecting in these circumstances yielded hundreds of new coral and other coelenterate species as well as an abundance of evidence to support the subsidence theory.[12]

Dana posited that the planet had gradually cooled and contracted, causing the crust to crumple like the skin of an old apple into mountain ranges and other features. In the context of this "shrinking earth" idea, subsidence of Pacific islands was due mainly to the cooling and contraction of the ocean floor. Dana included his theories in his U.S. Ex. Ex. reports *Zoophytes* (1846) and *Geology* (1849), and Darwin read the latter with glee. "To begin with a modest speech," he confessed in a letter to his friend Charles Lyell, "*I am astonished at my own accuracy!!* If I were to rewrite now my Coral book there is hardly a sentence I shd have to alter. . . . Considering how infinitely more he [Dana] saw of Coral Reefs than I did, this is wonderfully satisfactory to me. . . . He treats me most courteously.—There now, my vanity is pretty well satisfied."[13]

A Sublime and Awful Spectacle

Dana's Pacific travels also gave him greater opportunity than any previous naturalist had enjoyed to study and theorize about volcanoes. On Hawaii, where a missionary would later introduce him to the islanders as *kahuna wawahi pohaku,* the "rock-rending sorcerer," Dana visited the Kilauea volcano, where he "descended to the lowest depths, wandered over the heated lavas, through the hot vapors and sulphurous gases, and reached one of the boiling pools," he wrote to a friend in

The scientists with the U.S. Exploring Expedition brought back thousands of species new to science, including this Fiji coral (illustrated by expedition artist Alfred Agate), from Dana's *Zoophytes*.

1840. "The surface was in constant motion, throwing up small jets six or eight feet, which fell around the sides of the pool. There was no explosion, and only a dull grumbling sound." He remained that night, when "the deep red glow of the boiling lake, reflected by the walls of the crater, and lighting up the canopy of clouds . . . made a most sublime and awful spectacle."[14]

Visiting volcanoes across the breadth of the Pacific, he pieced together broader patterns. The conventional view, voiced by Darwin, Charles Lyell, and almost everybody else, was that the rugged Pacific island mountains had gotten their steep slopes and dramatic gorges from the action of waves and currents in the course of rising up from the sea. This "marine erosion theory" was also supposedly a partial explanation for the subsidence of the islands around which coral reefs had formed. But as he struggled up the steep gorges of Mount Aorai on Tahiti, Dana realized instead that torrential rains and mountain streams had carved out the gorges and done the literal dirty work of sweeping soil and rocks down into the valleys. "If we remember that these mountain streams at times increase their violence a million fold when the rains swell the waters to a flood," Dana wrote in *Geology,* "all incredulity on this point must be removed."[15]

"Dana was offering a geological explanation," according to historian David Igler, "for the romantic paintings and sketches of jagged Pacific coastlines and peaks going back to Capt. James Cook's first voyage. Three generations of Pacific explorers had pondered the natural forces creating those iconic Pacific scenes, to which Dana answered: 'erosion.'" Dana was also "eerily accurate," according to Igler, in describing an underwater chain of collapsed volcanic islands that stretched beyond the surviving Hawaiian islands for another fifteen hundred miles to the northwest.[16]

Dana was also the first scientist to describe what modern geologists call the Pacific "ring of fire." In an 1847 article with the ambitious title "Origin of the Grand Outline Features of the Earth," he reported that the Pacific Ocean was "nearly encircled" by "a grand volcanic border." It ran from New Zealand across to Java, then up in an arc along the East Asian coastline, across the Aleutian Archipelago, and down the Pacific coast of the Americas. This led him to suspect the existence of "some universal cause" at work, in the Pacific and elsewhere, through which "the very framework of the globe has received its characteristic features."[17] But Dana missed the real cause, mainly because he thought that the continents had formed and taken up their permanent positions during the earliest history of the planet, as it cooled and contracted. His "permanence theory" became highly influential, particularly in North America. It would be another century before iconoclastic young geologists demonstrated the true universal cause beyond rea-

Dana's hammer, compass, contact goniometer (for measuring crystal angles), and his personal copy of his *System of Mineralogy*. Early in his career, he built glass models of crystals like these as teaching aids.

sonable doubt: vast tectonic plates colliding and sliding across one another in the mantle of the Earth were the source of the fireworks Dana had described so vividly on the Pacific Rim.

Even with such missteps, the breadth and daring of Dana's thinking made him a key figure in the development of the Earth sciences. "As much as anyone, Dana transformed geology into a truly historical science," Michael L. Prendergast wrote in his doctoral thesis on Dana's life and thought. It was through Dana's *Manual of Geology* "that geologists learned to speak of Eras, Epochs, and Periods—divisions that were defined by events rather than by types of rocks."[18]

Faith, Friendship, and Science

Many of Dana's big ideas—his mistaken belief in the permanence of continents and oceans, his revelation about the cumulative effects of erosion over time, and his demonstration that small animals could construct vast coral reefs by a slow process of accretion—fit with the nineteenth-century geological thinking that the natural processes familiar in the present day—not just great catastrophes or acts of God—could explain major geological changes in the distant past. This "uniformitarian" thinking was "liberating geologists from the bonds of miraculous intervention," according to historian William Stanton.[19] It was also the necessary preamble to the idea of evolution by natural selection, in which familiar forces like disease, starvation, and predation could transform species, too. But uniformitarian thinking didn't liberate Dana, nor did he want it to.

Writing at the end of the nineteenth century, biographer Daniel Coit Gilman emphasized Dana's evolutionary perspective: "It is plain to see now that his mind was so saturated with the idea of evolution and his mode of thought so determined by evolution methods that he was bound by philosophic consistency to reach eventually a true evolution point of view in the case of the organic kingdom as well as in that of the earth."[20]

But in his biological work, Dana continued for years to believe that species resulted from special acts of creation by God as part of his benevolent plan for the Earth. His own meticulous work with species and his characteristic openness to new scientific evidence must have raised troubling questions about this belief. Even so, he clung to his profound religious faith. The painful struggle to keep these two contradictory beliefs intact was more poignant because it played out in correspondence with his friend Charles Darwin.

Dana was one of the first people in whom Darwin confided about the book he was writing. "I am working very hard at my subject of the variation & origin of species," Darwin wrote in September 1856. It might take a few more years to publish, Darwin thought, "but whenever I do the first copy shall be sent to you." He thought pigeon breeds offered "the most wonderful case" of how a variety of forms could descend from the same wild stock. "You will be rather indignant," Darwin wrote, "at hearing that I am becoming, indeed I shd say have become, sceptical on the permanent immutability of species: I groan when I make such a confession, for I shall have little sympathy from those, whose sympathy I alone value.—But anyhow I feel sure that you will give me credit for not having come to so heterodox a conclusion, without much deliberation. How (I think) species become changed I shall explain in my Book." In December, Dana wrote back answering a taxonomic question Darwin had posed and offering assurance of his broad trust in Darwin's science: "Believe me always glad to aid you in any way within my power; for I believe there is real truth in the results of your labors, and the best of foundations for general laws or principles."[21]

True to his word, Darwin sent Dana one of the first copies of *On the Origin of Species* in November 1859, not hoping "to convert "him but merely to persuade him "that more can be said in favour of the mutability of species, than is at first apparent."[22] But Dana had now suffered a nervous collapse of unknown origin. It forced him to cut back on his enormous workload—and spared him indefinitely from having to read Darwin's great book.

Darwin soon heard, as he put it in a note to Lyell, that Dana was "quite disabled in his head," and he wrote Dana that he was "most truly and deeply grieved" at the news. "For years I have been in your state, that an hour's conversation worked me up to that degree that I wished myself dead," he added. Dana and his wife were traveling in Europe on sabbatical in an effort to aid his recovery, and Darwin asked them to "spare us a week" at Down, promising he would be first to withdraw from any weighty conversation.[23] But the Danas avoided England.

When Darwin wrote again in 1863, Dana still had not read *On the Origin of Species.* But he had regained some of his former health, despite occasional relapses, and neither his nature nor his position as editor of the *American Journal of Science* would permit him to ignore new evidence or feign indifference to it. This was still the Dana for whom "not to change with the advance of science" was that dreadful thing, "persistence in error." He tried for a time to dispute evolutionary thinking and preserve his bedrock faith in a divine plan of creation. He wrote to Darwin in 1863 noting the absence in most cases of "those transitions by small differences" from one species to another "required by such a theory," though he acknowledged that this might be because "geology does not yet afford the facts required."[24] But his reasoning was often convoluted, and Dana must have known it. Moreover, O. C. Marsh's prodigious fossil collecting in the American West was beginning in the early 1870s to demonstrate just such "transitions by small differences," and this evidence was coming to Dana's doorstep at Yale.

Dana's first belated acknowledgment of Darwinian evolution came in the 1874 edition of his *Manual of Geology,* where his list of ideas "most likely to be sustained by further research," if not yet proven, included this sentence: "The evolution of the system of life went forward through derivation of species from species, according to natural methods not yet clearly understood, and with few occasions for super-natural intervention." He maintained, however, that the human species still required *"the special act of a Being above nature."*[25]

The rest of Dana's life "was a progressive surrender to Darwinism," as one biographer has phrased it. It culminated in the edition of his *Manual of Geology* published in 1895. There he conceded that his earlier speculative statements about the origin of species were not "in accord with the author's present judgment." Instead, "the evidence in favor of evolution by variation is now regarded as essentially complete." At the same time, he reiterated that "nature exists through the will and ever-acting power of the Divine Being," and that "the whole universe is not merely dependent on, but actually is, the Will of one Supreme Intelligence." Science and religion were effectively separate but equal. In a devastating essay a century afterward, the paleontologist Stephen Jay Gould posed the question: "What is the crucial difference between Darwin's transcendent greatness and Dana's merely ordinary greatness?" Dana, he answered, "could not, or dared not, abandon the traditional hope and succor of centuries: 'Rock of Ages, cleft for me, / Let me hide myself in thee.'"[26]

Dana retired from Yale in 1894, aged eighty-one, remarking that geology had left him no time until then to think of old age. He noted that he was, as always, "ready for the summons, and rejoicing in the prospect beyond time," but until then, as always, "I work and work." Revisions on that 1895 *Manual of Geology* filled his days and at times his nights, too. He told his son of a dream in which he tried to induce a New Haven neighbor with a big property "to allow me to locate there one of the largest of volcanoes. He thought a small one would do. So Geology keeps control." On April 12, 1895, he wrote a letter to a colleague about how the wind affects the way dust becomes deposited in valley plains.[27] That night, his heartbeat became irregular and two days later, Dana died quietly, in the expectation of eternal life.

CHAPTER 6

A Rumor of War

The nine-tenths, when attained, were only an additional stimulus for securing the remaining one-tenth.

CHARLES E. BEECHER on O. C. Marsh as a collector

In the beginning they were friends. It was Berlin in 1863, and both men were footloose in Europe, buffered from the horrors of the Civil War, browsing among museum specimens as they pursued their education in the natural sciences. They had plenty in common—both had lost their mothers early in life and were tended for a time by an aunt, and both had endured unwanted apprenticeships as farmers at the behest of overbearing fathers. But it was their science, already converging on the same topics in paleontology, that drew them together. They continued to meet and correspond after their return to the United States, as both began to advance in the study of prehistoric life. Edward Drinker Cope named a species, *Ptyonius marshii,* after Othniel Charles Marsh. Marsh returned the compliment with *Mosasaurus copeanus* that same year, and he thanked Cope by name for his helpful suggestions on a paper about amphibians, which were a Cope specialty.[1] There was nothing to suggest that the two would in time rank among the most bitter and vindictive enemies in the history of science.

Cope came from a wealthy Quaker family on the outskirts of Philadelphia. He was frail and undersized as a boy but animated and highly imaginative, with what a former playmate later recalled as "a bright and merry way" about him.[2] From the earliest age, he showed a keen interest in the natural world and a talent for sketching what he saw. As an eight-year-old visiting Philadelphia's Academy of Natural Sciences, he came away with a drawing of the entire skeleton of an ichthyosaur. His education consisted of careful tutoring by his father and a few years of formal

Before their friendship went horribly wrong, Edward Drinker Cope named a species of fossil amphibian after Marsh, and Marsh in turn named this marine reptile *Mosasaurus copeanus* to honor Cope.

schooling up to the age of sixteen. He never went to college, apart from a brief stint being tutored by the eminent paleontologist Joseph Leidy at the University of Pennsylvania, but he published his first scientific paper at the age of nineteen, and continued to publish prodigiously and often in haste for the rest of his life.

On his return to the United States in 1864, Cope obtained a position teaching natural science at Haverford College, which his grandfather had cofounded. A cousin in the administration, also a cofounder, pushed for his appointment. "Rather after the fashion of creating Renaissance cardinals," the historian Jane Pierce Davidson wrote, "Cope was literally hired one day, and made an A.M. [given his master's degree] the next."[3] That is, his beginnings depended, like Marsh's, on being from the right sort of family and on that family's readiness to wield its philanthropic influence on his behalf.

The burdens of teaching proved, however, to be too much of a distraction from research, and in 1868, Cope moved with his wife and their young daughter to Haddonfield, New Jersey, where he built a substantial house in the Gothic Revival style. He was drawn there mainly by the marl beds, lime-rich mudstone, left behind roughly 66 million years ago by a shallow Cretaceous sea. The marl was being quarried then for fertilizer, and one of the marl pits had already produced a headless but otherwise nearly complete dinosaur skeleton, named *Hadrosaurus foulkii* by Joseph Leidy. In 1866, at the invitation of a marl pit superintendent curious about unusual fossils that had turned up, Cope had visited and come away with a huge *Tyrannosaurus*-like dinosaur which, he noted with predatory glee, "was the devourer and destroyer of Leidy's *Hadrosaurus,* and of all else it could lay claws on."[4] He described it the same month he first saw it, and the name he gave it, *Laelaps aquilunguis,* suggested something of his inventive nature. The genus *Laelaps* came from the Greek "like the wind" and was also the name of the inescapable hunting dog in Greek myth that Zeus ultimately turned to stone in mid-leap. The species name *aquilunguis* meant "having claws like an eagle." Later, when the rivalry between them was raging, Marsh let the wind out of Cope's great find by noting that the name *Laelaps* had been preempted by a genus of parasitic mites. He replaced it with a name that was technically accurate, meaning "tearing lizard," but about as inspiring as an infected wound: *Dryptosaurus.*

In March 1868, though, Cope and Marsh were still friends, and they spent more than a week together tramping in snow and rain through the South Jersey marl country. Together, Cope reported, "we have procured three new species of Saurians." Marsh in turn invited Cope to visit New Haven, and as late as January 1870, Cope was still signing his letters to Marsh "I remain with much regard thy friend."[5]

For a time, they shared a delight in their science. Cope's instinctive grasp of ecology enabled him to see vanished worlds as if in a vision. "When he began to speak of the wonderful animals of the earth, those long ago and those of today," the bone hunter Charles H. Sternberg recalled, "so absorbed did he become in his subject that he talked on as if to himself, looking straight ahead and rarely turning toward me, while I listened entranced." Marsh likewise took an almost childlike pleasure in discovery. "Every new thing in his own sphere of investigation which revealed itself—everything which had in it the promise of a revelation—gave him happiness and stirred him to fresh activity," Yale president Timothy Dwight (sometimes called "the younger") later recalled. "When he had made it his own, and found it of true value, he hastened with joyful ardor to relate his good fortune to his friends, as if he had possessed himself of a hidden treasure . . . and thus brought the listener, for the time at least, into sympathy with his delight." It is tempting to imagine that Cope and Marsh talked like that to each other during the years of their friendship.[6]

But they were also very different men. Cope frequently seemed to suffer from too much imagination, and it was the underlying cause of his frantic working style. "If I know myself," he wrote to his father, in a solipsistically phrased 1864 letter from London, "I need every possible aid to distract myself from myself, and if I do not have it my health suffers." Without work to occupy his mind, he did not know how he would end up, "but I do not much doubt, in insanity." Being confined to

Arthur Lakes at first sent fossils to both rival paleontologists, but when Marsh made a preemptive deal, Cope was obliged to forward these teeth from a meat-eating dinosaur on to Marsh in New Haven.

one subject for too long, he added, "is ruinous. Existence becomes a burden; little occurrences are magnified into great griefs, etc."[7] This knack for perceiving slights and nursing them beyond all reason would be his undoing.

Marsh was at least as difficult a personality, as even those who admired him admitted. His landlady when he was a young man said it was "like running against a pitchfork to get acquainted with him." But her daughter, then a little girl, recalled him carrying her on his shoulders and delighting in showing her his minerals. He could be a deft and entertaining storyteller, but offstage he often spoke with "peculiar hesitations and interjections," as if lost in a conversation with himself or with his fossils. Though he was adept at striking up impromptu friendships with "those whom it was most desirable to know," as Timothy Dwight put it, he often neglected to communicate his thinking to students, assistants, and others down the social order. He could be socially inept, sometimes appallingly so, as when he wrote to the husband of his favorite niece, Julia, on the birth of their child. "I suppose it is my cousinly duty to offer my congratulations," he began unpromisingly. Then in a rush, he added, "Of course Julia thinks it the most beautiful baby in the world, but I would not give the fossil baby I got out of the Newark Mound for a half a dozen such." He was, he admitted, "a horrid old bachelor."[8]

Some writers have theorized—in the absence of evidence—that he remained a bachelor because he was a homosexual, or alternatively because "he would never be content with anything less than a 'collection' of wives." The more likely explanation is that in his "soul's deepest self," as Dwight put it, there was "a solitariness," an innate turning away from intimacy.[9] Were he alive today, he would no doubt be tagged with that overworked diagnosis "somewhere on the autism spectrum."

"People were disposed to laugh at his unusual ways rather than to observe the sterling qualities which lay beneath them," George Bird Grinnell, who worked with Marsh for a decade as a student and assistant, later wrote. "Marsh was a peculiar man and did not often show his real self to those with whom he casually came in contact." He lived alone, had few relatives, "and was really attached to very few people. Hence, in great measure his thoughts were about himself." He could be "somewhat selfish"—Grinnell scratched in that "somewhat" as an afterthought—though "most kindhearted" where his own interests were not involved.[10]

But should anyone trespass on those interests, Marsh "met all opposition with power and fearlessness," according to another former student and later colleague, the paleontologist Charles E. Beecher. His collections were his obsession, and Marsh "entered every field of acquisition with the dominating ambition to obtain everything there was in it, and leave not a single scrap behind. . . . The nine-tenths, when attained, were only an additional stimulus for securing the remaining one-tenth."[11]

It was probably inevitable that the early friendship between two such men as Marsh and Cope, the one inclined to giving slights, the other to taking offense at them, would end badly. But exactly when and where the terrible breach began became, like everything else between them, a source of heated debate. Cope later said he was infuriated to learn, sometime after the South Jersey trip, that Marsh had made a deal behind his back to monopolize future specimens from the owners of marl pits where Cope had until then collected. For him, this was tantamount to stealing another man's claim. But another incident that year may also have left Cope nursing his "wounded vanity," as Marsh later contended.[12]

The remains of a huge marine reptile had reached Cope from an army surgeon at Fort Wallace in western Kansas, and Cope had characteristically rushed it into print under the name *Elasmosaurus platyurus.* A full description, published in 1869, included a handsome illustration of a creature with a short neck, a wide, finned body, and an elongated, serpentine tail, which Cope thought served as a means of locomotion.[13]

Marsh later claimed that Cope showed him the skeleton laid out as he had arranged it at the Academy of Natural Sciences of Philadelphia and that Marsh had "suggested to him gently that he had the whole thing wrong end foremost." It was the neck that was long, not the tail. Mistakenly putting it the other way around might have resulted from Cope's prior work on lizards and amphibians, which tend to have long tails. "His indignation was great," Marsh wrote, "and he asserted in strong language that he had studied the animal for many months and ought to at least know one end from the other."[14]

Marsh's idea of suggesting "gently" may well have come off as a patronizing Ivy League put-down, especially given his tendency to tactlessness and the imbalance between the two men in academic credentials. That was certainly the tone Marsh took when he twitted Cope in print about the *Elasmosaurus* blunder as the conflict between them escalated in 1873. But some researchers have argued persuasively that it was Cope's former mentor Joseph Leidy who pointed out the error in the first place, not Marsh at all. That's what Cope seemed to be suggesting in a letter to Marsh in March 1870: "Once more I got into error by following Leidy," who had made a similar reversal on another New Jersey plesiosaur years earlier.[15] He wondered if Leidy would admit his own mistake to the Academy of Natural Sciences, but Leidy had in fact already done so in person and soon followed up in print. Humiliated, Cope tried to get back every copy of his published article with

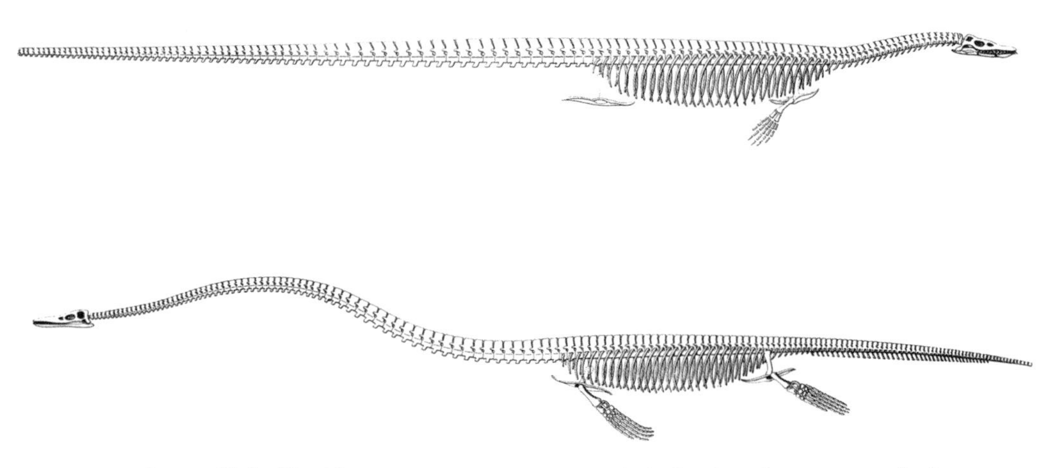

Cope published his *Elasmosaurus* reconstruction with the head on the wrong end (*top*), correcting it in a later edition (*bottom*). That mistake would become fodder in the infamous Bone Wars.

the backwards *Elasmosaurus,* and paid to print a revised version in which the illustration showed neck and tail properly reversed.

One way or another—by claim-jumping or by anatomical insults—the friendship between Cope and Marsh soon came to an end. The poison of their developing rivalry would bleed out across the wild countryside of the American West—and in newspaper headlines—for decades to come.

CHAPTER 7

The Marsh Expeditions

There is no collection of fossil vertebrates in existence, which can be compared with it.
T. H. HUXLEY on visiting the Peabody Museum, 1876

The 1870 expedition drew to a close late that November with a near disaster. The Yale party had headed out from Fort Wallace in western Kansas with a military escort to explore along the Smoky Hill River. Cretaceous beds in the Niobrara chalk bluffs there would turn out to contain some of the richest prizes for Marsh's collection, that year and beyond. The men pitched their tents "under the lee of a high bluff near the river," Marsh wrote, "in order to protect ourselves as far as possible from the cold northerly winds that then prevailed." The bluff also sheltered them from attack on that side, with the tents and animals huddled below and the wagons drawn up in a half circle around them. "Thus secure," Marsh wrote, they were sleeping one cold, windy night when "we were suddenly aroused by fearful noises of clashing wagons, snorting horses, and falling tents, as if an earthquake were in progress directly under our camp"—or, more likely, as if Indians were attacking. They stumbled out of their tents, imagining death by arrows winging silently out of the dark. It turned out instead that the animals had stampeded, panicked by a coyote leaping frantically down from the bluff into the middle of the camp to get at the buffalo meat hanging by the cooking tent. When relative calm returned, one of the soldiers mounted an unpanicked but obstinate donkey named Crazy Jane and headed back fifteen miles to the fort for help.[1]

The stampeding animals got there first, "with their broken halters and dragging lariats," spreading the panic to the fort commander, who exclaimed, "Great God, the Indians have jumped the Professor!" A company of soldiers promptly rode out to the rescue. Marsh and his students meanwhile went about their usual business of hunting for fossils and were just returning to camp at sunset when the rescue party arrived. "The troops appeared more disappointed at losing the expected fight than gratified at our safety," one of the students wrote. It was Thanksgiving Day, and explanations having been given and the coyote having failed to carry away the buffalo meat, the Yale party invited the army officers to join a proper feast. Celebratory speeches ensued, "Yale songs were sung by our party, and western stories were told by the army officers," Marsh wrote. "The November wind howled through our camp, and the coyotes, sniffing the feast, serenaded us from the orchestra bluff above, but we heeded them not, for we were safe and happy."[2]

The chalk bluffs produced numerous fossil remains of mosasaurs, huge sea serpents like monitor lizards, with paddles instead of feet. But what seemed to be the great find of the season was a small, hollow bone fragment, described by Marsh as about six inches long and an inch in diameter, "with one end perfect and containing a peculiar joint that I had never seen before." Marsh thought at first that it might be "the tibia of a gigantic bird." But careful study back in New Haven, where the party arrived a week before Christmas, revealed that it was a knuckle—specifically, a knuckle of the finger that extends all the way out along the leading edge of the wing of a pterodactyl. But all previous specimens of these "flying dragons" had been relatively small, "not one-twentieth the size or one-hundredth the bulk this bone indicated." Marsh plucked up his courage and consciously modeled himself after the great French anatomist Georges Cuvier, who had coined the name "Ptero-Dactyle" (literally "winged finger") in 1809. Cuvier was revered for an almost magical ability to take a fragment of bone from the distant past and, by a series of inferences, often based on his deep knowledge of modern animal anatomy, reconstruct an entire prehistoric species. In that spirit, Marsh published his description with the bold assertion that pterodactyls with a twenty-foot wingspan had once ruled the skies over Kansas.[3]

Doubt about this conjecture haunted his sleep that winter, and on his second Yale expedition the following June Marsh headed straight back to Kansas, hoping to uncover further evidence. Working with his hunting knife "and a small brush made from buffalo grass," he soon found other pieces of his original specimen. That year's party of ten students unearthed an assortment of other individuals. Marsh was "soon able to determine that my calculations based on the fragments were essentially correct, and that this first found American dragon was fully as large as my fancy had painted him." A wingspan of up to twenty-five feet might even be possible—large enough, he added, a bit too imaginatively, "to carry off a man."[4]

The real prize from Kansas that summer, though, was a "gigantic swimming bird" or, as he later put it, "essentially a carnivorous, swimming Ostrich," five or six feet in height. Marsh named it *Hesperornis regalis,* "the ruling western bird."[5] Its true significance became evident only the following year, when T. H. Russell, a student on Marsh's 1872 expedition, turned up another individual with a nearly complete skull. Scientists already suspected that early birds had reptilian traits because of the sensational discovery of *Archaeopteryx* in 1861, soon after publication of Darwin's *On the Origin of Species.* But because that first *Archaeopteryx* specimen was incomplete, no one knew that the 150-million-year-old *urvogel,* or original bird, had teeth. The Marsh-Russell *Hesperornis* specimen from 1872 would thus have been a stunning find all by itself: not only did it have formidable rows of reptilelike teeth, apparently for grasping prey underwater, but it was also a much more modern bird than *Archaeopteryx,* from just 80 million years in the past. That summer of 1872 would produce a second, similarly sensational fossil bird under circumstances that were now complicated by the developing rivalry between Cope and Marsh.

The Battle of Bridger Basin

That rivalry had surfaced briefly in western Kansas the previous September 1871, when Cope arrived to work in the Smoky Hill River country near Fort Wallace just a few weeks after Marsh and his party departed. The shale hillsides and stream banks there had already given Cope his *Elasmosaurus* skeleton as well as other specimens sent to him at the Academy of Natural Sciences by a local geologist and educator named Benjamin Mudge. Cope had every right to be there, but he was no doubt also attracted by Marsh's highly publicized discoveries. "Marsh has been doing a great deal I find," Cope reported to his wife, "but has left more for me." His expedition lacked the protection of a military guard, but reports of Cheyenne war parties in the area did not dissuade him. He learned enough of Marsh's route from the soldiers at Fort Wallace to at least "save my going over the same ground." They also told him stories of what Marsh had found, he wrote, "and I can now tell my own stories, which for the time I have been here are not bad." While Marsh's "flying dragon" had been impressive enough, Cope could soon boast of an even bigger pterodactyl, with wings spanning almost twenty-five feet.[6]

The rivalry between the two men had been simmering at the level of scholarly sniping in correspondence and in academic papers.[7] But in the following summer of 1872, it spilled into open conflict when Cope, Leidy, and Marsh all showed up in Wyoming's fossil-rich Bridger Basin. Leidy had been publishing species descriptions since 1869 based on specimens sent to the academy by amateur fossil

Marsh (*seated, third from left*) with his 1871 expedition crew at a site near Salt Lake City.

hunters at Fort Bridger, and Cope had begun to do so in 1871. Marsh had meanwhile been conducting fieldwork there with his students since 1870 and had, predictably, developed a sense of ownership.

Cope arrived at Fort Bridger in the third week of June 1872, too early in the season to do much collecting. The way he planted himself there, when there was an entire West of other sites waiting to be discovered and explored, smacked of spoiling for a fight. He had recently obtained a position as an unsalaried scientist with the U.S. Geological and Geographical Survey under Ferdinand V. Hayden. But when Hayden departed with all of Fort Bridger's available horses, pack mules, saddles, and tack to the Yellowstone region, Cope opted not to go along. Instead, he sent a letter to General Edward Ord, then commanding the Department of the Platte, advising "that I have charge of the department of Palaeontology for the survey of 1872." Having detailed his needs, he added: "I submit my hope that the requisite outfit in men and teams may be found elsewhere, as you shall see fit. You can inform me at your pleasure what can be done."[8] But Cope lacked Marsh's gift for knowing the right people and enlisting them at every turn in his cause. It was General Ord's pleasure to do nothing.

Strong Words

Leidy, a shy, retiring figure, was appalled by the bad behavior of the two rivals and by their readiness to offer money for specimens that "used to come to me for nothing." He soon withdrew entirely from paleontology. But Cope and Marsh continued to trade accusations and admonishments by mail. In a letter early in 1873, Marsh accused Cope of stealing his specimens and declared that he was so enraged by the hiring away of one of his trained collectors that "I should have 'gone for you,' not with pistols or fists, but in print." Cope responded with his own accusations of unsavory acts, addressing Marsh as "O See," a nickname supposedly bestowed on him by New Haven women for his pomposity. He also mocked him, in a letter to his father, as "the learned professor of Copeology."[15]

Their personal letters to each other stopped abruptly in February 1873, but then the two carried on their war in a series of published articles, largely focused on dates of publication and supposed errors. The debate "waxed very warm," the *Chicago Inter-Ocean* reported in July 1873, "so much so that these great guns of science, who can comprehend the profundities of evolution and reconstruct all the dead creations of the earth, came down into a quarrel quite as undignified as one that might be waged between a couple of very belligerent city or country editors. Cope thus far has the last word, which amounts to simply this: that he will have nothing more to do with Professor Marsh, whose 'recklessness of assertion, erroneousness of statement, and incapacity of comprehending,' etc., 'render further discussion unnecessary.' Next month Professor M. will doubtless retort: 'You're another,' and there it will end."[16]

Sadly, though, the paleontological warfare did not end. When the editors of the *American Naturalist* had had enough, they told the two combatants they could carry on their fight in paid ads at the back of the journal, as though, David Rains Wallace wrote, "they were sending bad boys to the cloakroom."[17]

The rivalry took over a portion of each man's mind and began to fester. On an 1873 expedition, Cope wrote of collapsing from overwork and infection: "During my fever I had terrible visions and dreams, and saw multitudes of persons, all speaking ill of something and frustrating my attempts to sleep." After calming himself with a mixture of opium, quinine, and belladonna, he was able to report, "My nights are positively happy under the influence of an opiate."[18]

Marsh meanwhile became more secretive and withdrawn. "We found it very difficult to get any information from Professor Marsh on what we were doing," wrote Henry W. Farnam, a member of the 1873 expedition and later an economics professor at Yale. "I cannot recall that he ever gave us even a cursory lecture on the

Marsh (*standing, center*) and his crew of Yale men, armed to the teeth. *Upper right,* the lieutenant with a U.S. Army escort for this 1872 expedition.

geological formations on which we were working or the possible significance of what we were finding. If we asked him questions, he was very apt to give a few of his characteristic grunts."[19]

In a letter to his father, Cope wrote, "Cuvier never had as important and curious forms to illustrate, and it seems incredible that our Societies should be mean about it." Henry Fairfield Osborn called it "Cope's truest comment on the struggle with Marsh." But the personal application of those words was lost on the two combatants. It wasn't "our Societies" that were guilty of meanness but the two men, who should have felt blessed to be making some of the greatest discoveries in the history of paleontology.[20]

Birds and Reptiles

Probably as a result of the perceived trespass on his territory, Marsh sent a note in September 1872 to a fossil collector back in Kansas, the same Benjamin Mudge who had been shipping specimens to Cope at the Academy of Natural Sciences. Only the year before, Marsh had refused to allow Mudge, a geologist at Kansas State Agricultural College, to accompany his 1871 expedition in the Smoky Hill region because of his connection to Cope. But now Marsh offered to identify "any reptilian or bird remains without expense" and promised Mudge "full credit" for their discovery. Cope had treated Mudge well enough. But Mudge had known Marsh back in Massachusetts in the mid-1850s as a fellow mineralogist and of course also as a Peabody, and he must have sensed a greater opportunity there. In any case, he took a package he was about to ship and changed the label from Philadelphia to New Haven.[21] Thus Cope lost out on one of the great finds in evolutionary history.

On first examining the shipment soon after, Marsh thought a couple of toothy jaws belonged to a new reptile, which he named *Colonosaurus mudgei.* Further work scraping away the stone in which the fossils were embedded revealed, however, that the jawbones actually belonged to another bird with teeth, a ternlike creature Marsh named *Ichthyornis dispar.* The discovery of both *Ichthyornis* and *Hesperornis,* he announced triumphantly in 1873, "does much to break down the old distinction between Birds and Reptiles, which the *Archaeopteryx* has so materi-

The toothed bird *Ichthyornis dispar,* together with the much larger *Hesperornis,* solidified the link between reptiles and birds. When the museum first displayed them, some thought they "savored too much of evolution" and ought to remain hidden from view.

ally diminished." He thought that *Archaeopteryx* would ultimately also be proved to have teeth, a prediction soon proved correct.[22]

Cope was undoubtedly angered when he discovered the importance of the find that had so narrowly eluded him. Others found the idea of toothed birds disturbing on more philosophical grounds. When *Ichthyornis* went on display at the Peabody Museum, visitors "far more religiously zealous than wise" urged Marsh to conceal it from the public, according to Erwin H. Barbour, a paleontologist under Marsh in the 1880s, "because it savored too much of evolution." They admitted that it was genuine — it was right there in front of their eyes in the display cabinet. That was precisely the problem. They proposed that "an opaque curtain be so arranged as to be drawn over the specimen to conceal" the awful truth. That truth, as T. H. Huxley later declared, was that Marsh's toothed birds had "completed the series of transitional forms between birds and reptiles" and moved Darwin's evolutionary conjecture about these transitional forms "from the region of hypothesis to that of demonstrable fact."[23]

"The Last Toe Wanting"

The lack of such transitional fossils in the decade after Darwin published *On the Origin of Species* had caused skeptics to assert loudly that evolutionists could produce no evidence of how ancient species evolved into modern ones. Huxley responded with what he thought was just such a sequence, for a species almost as familiar in the Victorian era as the dog. Building on recent work with fossils from Europe by the Russian paleontologist Vladimir Kowalevsky, he described an evolutionary sequence leading in four steps, over roughly 50 million years, from the horselike *Palaeotherium* (literally "old beast") to *Equus*, the modern horse, then still the engine of Victorian society. Unfortunately, as one history later put it, this "sequence was not representative of the main line of horse evolution, because that story took place elsewhere."[24]

Huxley found this out for himself soon after, in August 1876, when he came to inspect O. C. Marsh's collection at the Peabody Museum. The two men had become acquainted during Marsh's visits in London and now, at the beginning of an American lecture tour, Huxley wanted to see for himself the discoveries he had heard so much about. He spent a week of "hard labor" at it, according to Marsh, paying particular attention to the horse specimens. He had already prepared a lecture on horse evolution based entirely on European specimens. "My own explorations had led me to conclusions quite different from his," Marsh later

When T. H. Huxley defended Darwinian evolutionary theory, key evidence came from the gorilla, a species discovered by a student of Silliman's, and from bird and horse specimens at the Peabody Museum.

recalled, "and *my* specimens seemed to me to prove conclusively that the horse originated in the New World and not in the Old, and that its genealogy must be worked out here. With some hesitation, I laid the whole matter frankly before Huxley, and he spent nearly two days going over my specimens with me, and testing each point I made."[25]

Marsh had already named numerous extinct horse species and five horse genera that are still recognized today. So when Huxley asked to see a specimen illustrating some point, or if he wanted a fossil to show a more generalized early form evolving into a more specialized later one, "Professor Marsh would simply turn to his assistant and bid him fetch box number so and so," Leonard Huxley, his father's biographer, later wrote. Huxley exclaimed, "I believe you are a magician; whatever I want, you just conjure it up." It wasn't magic, of course, but the result of Marsh's four student expeditions from 1870 to 1873 and the work of his hired collectors, all of them with their noses in the dirt, braving intense heat and bitter cold, relentless mosquitoes, rattlesnakes, grizzly bears, hungry coyotes, bad food, robbers, Indian war parties, and other perils in the cause of paleontology. Writing a few days later about what he had seen at the Peabody, Huxley remarked, "There is no collection of fossil vertebrates in existence, which can be compared with it." "The more I think of it," Huxley wrote Marsh from his

These fossil fragments from the "hatful of bones" collected for Marsh in 1868 made him realize that horses, long considered a European import, had once been indigenous to North America.

next stop in Newport, "the more clear it is that your great work is *the settlement of the pedigree of the horse.*" A gratified Marsh later recalled, "With the generosity of true greatness, he gave up his own opinions in the face of new truth, and took my conclusions."[26]

Huxley asked Marsh to prepare an illustration to help him as he revamped his lecture, and Marsh responded with what Stephen Jay Gould described more than a century later as "one of the most famous illustrations in the history of paleontology—the first pictorial pedigree of the horse."[27] Marsh lined up leg bones and molars of different North American species, increasing in size over time, from the 50-million-year-old *Orohippus,* with four toes on its front legs, on up to the modern horse with a single hoof. Marsh's illustration made its auspicious debut that September, when Huxley spoke at Chickering Hall in New York.

"By means of a diagram at the rear of the stage," the *New York Times* reported, Huxley "led his audience back with him through the chain of different equine types, beginning with the existing horse, and ending with the newly discovered specimens of Prof. Marsh, on which the hoof consists of four separate fingers, as in the hand of man. The lecturer added that in all probability, when the geological formation of the great North-west could be thoroughly examined, a still more ancient species will be found, with a fifth finger or thumb on the forehoof, and thus complete the series. Enough, however, had been discovered to demonstrate the truth of the evolution hypothesis—a truth which could not be shaken by the raising of side issues." The *New York Herald* headlined its coverage, "Horses with Fingers and Toes Discovered in America: The Last Toe Wanting."[28]

Huxley and Marsh had joked about this imagined proto-horse one evening at the end of their work in New Haven. They dubbed it *Eohippus,* the "dawn horse," and as they chatted, Huxley began sketching a fanciful five-toed horse on a sheet of brown paper. "That is my idea of *Eohippus,*" he remarked. After a moment, he added, "But he needs a rider," and he penciled in an equally fanciful primate riding bareback.

"The rider also must have a name," Marsh thought. "What shall we call him?"

"Call him '*Eohomo,*'" Huxley replied, whereupon Marsh inscribed both names with a swirling flourish at the bottom: *Eohomo* and *Eohippus*—that inseparable duo, the cow-ape and his nag, ambling together out of the sunrise of some improbably ancient past.[29]

Five months later, Marsh wrote Huxley to announce that he had found *Eohippus:* "I had him 'corralled' in the basement of our Museum when you were there, but he was so covered with Eocene mud, I did not know him from Orohippus. I promise you his grandfather in time for your next horse lecture if you will give me proper notice." Soon after, at a meeting of the American Association for the Advancement of Science, Marsh declared, "To doubt evolution to-day is to doubt science, and science is only another name for truth."[30]

The horse thus became one of the classic models of how evolution works, and for a century or more, textbooks and museum displays featured it as what Ernst Haeckel called "the parade horse of paleontology." These evolutionary accounts typically included updated versions of Marsh's illustration, which made the evolution of horses look, Stephen Jay Gould complained, "like a line of schoolchildren all pointed in one direction" and arranged according to size. It helped popularize the mistaken idea of an evolutionary ladder, with species progressing over time in a direct ascent toward some more perfect form. But in evolution, as in human families, genealogies tend to be less neat and orderly (and, sadly, less progressive) than we like to imagine. They are not ladders, Gould wrote, but "copiously branching bushes—and the history of horses is more lush and labyrinthine than most."[31] He credited the bush metaphor, which soon displaced the ladder in evolutionary discussions, to a 1951 account by George Gaylord Simpson of the American Mu-

After a long day at the Peabody Museum going over evidence for Marsh's account of horse evolution, Huxley sketched a "dawn horse" ridden by an even more imaginary "dawn man."

seum of Natural History. Ironically, Simpson had started out as a doctoral student at the Peabody Museum where, among other things, he studied Marsh's horses.

The paleontological writer Brian Switek recently pointed out another irony in this chronicle of equine evolution: Cope had a "dawn horse" of his own, though he called it *Protorohippus,* and while Marsh's *Eohippus* sat on a shelf for years after he described it, "Cope's employee J. L. Wortman uncovered the rest of Cope's specimen of *Eohippus* in 1880." The resulting reconstruction influenced Charles R. Knight, the great illustrator of extinct animals at the American Museum of Natural History. Through Knight, Cope's *Eohippus* specimen in turn became the model Rudolph Zallinger relied on when he was creating the "dawn horse" for his great *Age of Mammals* mural—on the walls of O. C. Marsh's Peabody Museum.[32]

The final irony of Marsh's discoveries in those years was that they would become evidence not just for Darwinian evolution but also for at least one form of religious belief. In the middle of their 1873 expedition, Marsh and his students had retreated to Salt Lake City for a much-needed rest and found themselves being feted as heroes by Brigham Young and other members of his Mormon Church. Young took a surprisingly detailed interest in fossil horses, interrogating Marsh closely about his discoveries. It soon came out that a critic in London had once debunked the *Book of Mormon* because it spoke of horses in prehistoric America when everyone knew that no such animals existed there before the Europeans arrived. "So it seems that while most theologians are regarding the developments of the natural sciences with fear and trembling," the *New-York Daily Tribune* reported, "the chiefs of the Mormon religion are prepared to hail the discoveries of palaeontology as an aid in establishing their peculiar beliefs . . . and thus Prof. Marsh, one of the warmest advocates of the development theory, is raised to the rank of a defender of the faith."[33]

This was the same Marsh, and these were some of the same fossils, that would lead another correspondent to write in 1880, "Your work on these old birds, and on the many fossil animals of North America, has afforded the best support to the theory of Evolution, which has appeared within the last twenty years." That letter was signed, "With cordial thanks, believe me, Yours very sincerely, Charles Darwin."[34]

CHAPTER 8

Professor M on the Warpath

He told the Great Father everything just as he promised he would, and I think he is the best white man I ever saw.

RED CLOUD, 1877

After his fourth student expedition in 1873, Marsh withdrew, for the most part, from directly participating in fieldwork. He was busy enough sorting through the tons of fossils he had already acquired and preparing papers for publication. Moreover, the threat of being scooped was now driving Marsh and Cope alike to establish priority for their discoveries, which meant getting them into print as quickly as possible—sometimes too quickly for the purposes of good science. Construction of the new Peabody Museum had at last also begun, making it impractical for Marsh to vanish for six months in the American West.

Finally, Marsh seems to have come to terms with the power of his inherited wealth to accomplish much of the dirty work of acquisition on his behalf. When news broke early in 1873, for instance, that a remarkable pterodactyl specimen (actually a kind of pterosaur, in modern terminology) had been discovered near Eichstätt, Germany, other museums asked for photographs and dithered about the price. Marsh simply swooped down, swift and predatory, with a six-word cable to a museum friend in Germany: "Buy Eichstätt pterodactyl for Yale College." The friend gulped at the asking price, almost $1,100 but, knowing his buyer, made the deal. *Rhamphorhynchus phyllurus,* as Marsh named it, was the first pterosaur ever found with clear impressions of the wing membranes, a key to understanding pterosaur flight, and it "was on its way to Yale," according to Marsh biographers Schuchert and LeVene, "before the other museums had received answers to their cautious requests."[1]

Oglala Sioux chief Red Cloud's war bonnet, purchased from his son.

Marsh snapped up this fossil (*Rhamphorhynchus phyllurus*) because it was the first pterosaur ever found with preserved wing membrane impressions.

But he missed out on acquiring this sensational specimen, depicted in a rubbing from the original, partly because the Germans were determined to keep it at home. It is now known as the Berlin *Archaeopteryx*.

In his travels around the West, Marsh had always been careful to connect with other collectors and he now used his wealth to deploy hired collectors across the region for the 1874 season, in place of students. Some of them had academic training, including Mudge, who became a full-time employee. Others were more like prospectors, "men accustomed to frontier work and to the somewhat sporadic sort of labor involved," according to Schuchert and LeVene, "men who could pick up a grubstake, a horse or two, and a rifle, and set off into the fields to stay until their provisions were exhausted; men to whom the regular arrival of a pay envelope was not an essential."[2] (A good thing, as Marsh was notoriously slow with wages.)

All of them had the wilderness equivalent of street smarts, a knack for surviving in the most extreme circumstances and with no hope that the cavalry was about to come riding to their rescue. They knew how to use their guns, of course, but also their wits. Benjamin Mudge, for instance, may have seemed like a quaint professorial figure waking up the camp each morning whistling his favorite tune, "'Tis Morn, the Lark Is Singing." But he had also once survived an encounter with a Sioux war party by grinning broadly, then popping his false teeth halfway out of his mouth and back again, a spectacle so bizarre the warriors "begged him to do again and again to their intense delight."[3]

Marsh and his collectors quickly developed a code language for telegraphing news without revealing the contents to potential spies. Cope was "Jones" or "B. Jones." "Ammunition" was money (aptly for almost anything Marsh undertook), and "health" was success. One result was this sentence of hieroglyphic inscrutability, telegraphed from western Kansas on April 22, 1877: "Send hundred ammunition health poor B. Jones nothing going south." Roughly translated, it meant: "Send a hundred dollars. We're not having much luck, but neither is Cope, who has now departed for Texas."[4]

Late in 1874, Marsh made his unexpected last hurrah as a field paleontologist. His knack for campfire camaraderie and the infectious excitement of the work he was engaged in had won him many friends in the military. But the series of invitations he received that year suggests that his scientific work may also have served, despite Marsh himself, as a useful cover for other agendas. In May, General Sheridan wrote Marsh about a planned expedition into the potentially fossil-rich territory of the Black Hills of Dakota by General George Armstrong Custer, noting that he would not allow "any outside people except yourself if you should desire to go." In June, Colonel T. H. Stanton in Cheyenne wrote to advise Marsh of the discovery of fossils south of the Black Hills, and Marsh, intrigued, wrote back to ask if someone there might be able to collect on his behalf. In July, Custer wrote repeating the invitation to join his excursion, and again Marsh had to decline. This time, though, he sent in his place George Bird Grinnell, a veteran of the 1870 expedition who was now back at Yale as a graduate student and assistant. Finally, in October, General Ord wrote to report that "a vast deposit of fossil remains of extinct marine and other animals has been discovered ten miles north of the Red Cloud Agency covering an area six miles square." Many of them were strewn across the surface in a kind of boneyard of extinct species. This was too much temptation for Marsh, who packed his bags and caught the next train to Cheyenne. He seems to have been unaware that he was entering what one historian has aptly called a "maelstrom of politics, prejudice, corruption, and violence."[5]

In the mid-1860s, the Oglala Sioux chief Red Cloud had waged a ferocious guerilla campaign against soldiers and settlers along the Bozeman Trail in the Powder River country of Wyoming. His warriors perfected the tactic of sending decoys to lure a military patrol into an ambush, leading to the 1866 incident known alternatively as the battle of the 100 Slain or the Fetterman Massacre. It began with a small party of Indians on a ridgeline mocking and even mooning a patrol led by Captain William J. Fetterman. Outraged, Fetterman foolishly ordered his command to charge straight into what turned out, on the other side of the ridge, to be a welcoming party of two thousand or more Sioux. All eighty-one soldiers died, and their bodies were scalped and otherwise mutilated. It was the most infamous military defeat on the Great Plains until the battle of the Little Big Horn a decade later. As a condition of peace, Red Cloud ultimately forced the military to withdraw from its three Powder River forts, which the Sioux then burned to the ground. That brought an end to Red Cloud's War, and by treaty in 1868, the U.S. government ceded all of the territory west of the Missouri River in what is now South Dakota, more than half the modern state, as a reservation "for the absolute and undisturbed use and occupation of the Indians."[6]

Three years later, ignoring this treaty, government representatives informed the Sioux of their intention to run the Northern Pacific Railroad just north of the supposedly permanent reservation, ensuring the arrival of new settlers. Among themselves, they concluded that the Indians had no power to resist. But just in case, General Sheridan was soon proposing construction of a fort in the Black Hills, sacred territory to the Indians, and in July 1874, Custer launched his planned expedition, effectively an invasion, with one thousand soldiers, a marching band, and two mining experts to substantiate rumors of gold. They met no resistance, and a sergeant later recalled his time in the Black Hills as a "prolonged picnic," with music. Custer was soon reporting, "I have upon my table 40 to 50 small particles of pure gold," adding, "Veins of what the geologists term gold-bearing quartz crop out on almost every hillside." Word quickly spread to the outside world, stirring up a new bout of gold fever. By the end of the following year, fifteen thousand prospectors would be at work in the Black Hills. The Sioux would remember Custer's expedition as the Trail of Thieves—and they would avenge it.[7]

With winter approaching, Sioux hunting parties, many of them hostile to military intrusion and angered by Indian Agency finagling, began crowding onto the reservation. An incident that October almost turned bloody when J. J. Saville, the Indian agent at the Red Cloud Agency near what is now the town of Crawford in northwestern Nebraska, ordered a flagpole to be erected in front of his stockade. An angry crowd soon assembled to protest the raising of an American flag on what was left of their nation. Braves, armed and dressed for a fight, jostled around the agency workers and drove them off with their war cries before hatcheting the flagpole to splinters. When a party of twenty-six soldiers rode in to restore peace, the crowd enveloped them, shouting and firing weapons in the air. The soldiers were barely able to back out with their lives.[8] Red Cloud himself was now also angrily complaining about the wretched quality of the food, blankets, and other necessities promised as part of the treaty. And that was assuming these essential supplies were delivered at all: the Indian chiefs were now dealing with Saville's refusal to supply anything unless the Indians first submitted to a census.

This was the feverish atmosphere into which an unwitting O. C. Marsh rode in early November, intent on entering Indian Territory to reach the reported fossil beds south of the Black Hills. The council of chiefs assumed he was just another gold-hungry intruder—who else would be foolish enough to head out into the frostbite cold of a Great Plains winter?—and refused to grant permission. A lengthy palaver ensued, with Marsh offering to hire Sioux helpers, and Sitting Bull demanding wages of $5 a day, roughly five times the going rate. Marsh ultimately agreed to $1.50 and also, more critically, offered to carry Red Cloud's complaints back to his many powerful friends in Washington, bringing the negotiations to a successful conclusion. Snow the next morning delayed Marsh's departure. When he showed up the following day accompanied by military brass eager to join the fossil hunt, together with the well-armed soldiers who would serve as their guards, the Indians thought this was more than they had bargained for. An angry crowd blocked their path, and when weapons appeared, Marsh and his party withdrew.

Another lengthy palaver followed, with Marsh providing a feast and gifts to the assembled chieftains. He delivered a speech and eventually won a reluctant permission to proceed. Rather than risk another change of heart, he waited only till the celebrants had returned to their tents and gone to sleep. Shortly after midnight, he and his party, minus the Indian guides, slipped quietly through the tepee villages and out into open country. It was worth the risk: "Nearly two tons of fossil bones were collected," the *American Journal of Science* reported, "most of them rare specimens, and many unknown to science," notably including the massive, rhinolike ungulates called Brontotheridae. "At one point these bones were heaped together in such numbers as to indicate that the animals lived in herds, and had been washed into this ancient lake by a freshet."[9]

Along with his fossils, Marsh carried home the samples Red Cloud had provided of Indian Agency supplies that were unfit for human consumption. He seems to have been intent on becoming the one white man to keep his word to the Sioux, and on an April 1875 trip to Washington, D.C., he took Red Cloud's samples to E. P. Smith, the commissioner for Indian Affairs. Just two years earlier, Smith had written a letter on Marsh's behalf to the Lakota chief Spotted Tail when Marsh was first considering an expedition into the Dakota Badlands. Now, though, Red Cloud's charges failed to elicit the expected outrage or surprise from Smith. So Marsh proceeded next day—as only he could—to President Ulysses S. Grant and repeated his story.

Accounts of Marsh's developing campaign began to appear in the press, which in the past had reported enthusiastically on waste and corruption in the Office of Indian Affairs and the Grant administration generally. The members of the Board of Indian Commissioners asked Marsh to repeat Red Cloud's allegations at what they seem to have considered a private meeting in New York, but a detailed account appeared in the *New York Tribune* two days later, written by a guest Marsh himself had brought along to the meeting. A heated exchange of letters in the press soon followed between Marsh and a defender of the Office of Indian Affairs, who implied that Red Cloud was playing Marsh for a fool.

At this point, Secretary of the Interior Columbus Delano roused himself from what was apparently a deep sleep and wrote to the board requesting an investigation of "reports put in circulation by a Mr. Marsh." The use of that indefinite article, "a" Mr. Marsh, instantly made Delano a laughingstock, Marsh by then being one of the most famous people in the nation, vastly more so than Delano himself. The *New York Tribune,* according to Schuchert and LeVene, "practically burst into flame at the insult."[10]

A Sioux delegation, including Red Cloud, soon came to make its case in Washington. But administration officials were less interested in investigating charges of systematic maltreatment than in persuading the Indians to sell the Black Hills and give up hunting rights in Nebraska and Kansas. Red Cloud now also proved to be ineffective as a spokesman, even seeming to back away from the complaints he had made via Marsh.

Nonetheless, the debate about mistreatment of the Indians became a topic of intense national interest over the next three weeks. As the drama unfolded, J. J. Saville, the Indian agent at the Red Cloud Agency, demanded a full investigation, presumably with the idea that it would exonerate him. Indian Commissioner Smith then asked that Marsh submit a formal statement of charges, and Marsh responded, almost overnight, with a detailed thirty-six-page pamphlet in which he implied that Delano and Smith were aware of what was happening under their watch and had done nothing to stop it. A cover letter addressed to President Grant included the word *corruption* twice and variations on *fraud* fourteen times.[11]

Marsh now came under intense personal attack, much of it apparently originating in the Interior Department itself. His critics accused him, according to Schuchert and LeVene, "(1) of raising all this fuss to get notoriety and turn an honest penny by writing long articles for the *Tribune;* (2) of being the catspaw of the *Tribune,* whose editor, Whitelaw Reid, had a personal grudge of long standing against Secretary Delano; (3) of being bribed to submit the complaints of Red Cloud; (4) of being 'impetuous but not very practical—a nervous, impulsive, credulous man, naturally combative'; and (5) of holding 'sculking (*sic*) interviews' with the Indians while they were in Washington." The combative part at least was true, but friends advised Marsh to hold his fire: "All you have to do now is to keep perfectly quiet and let 'em wriggle," a member of the board of the *New York Tribune* advised. Henry Ward Beecher's *Christian Union* expressed what seems to have been the general opinion in an editorial: "Professor Marsh has all the qualifications of a first-class witness: unimpeachable character; uncommon intelligence, the absence of any personal motive—the whole business being a distraction from the professional pursuits to which he is devoted—and excellent opportunities for personal observation" of the shady business in question.[12] Others who had had firsthand dealings with the corrupt "Indian Ring" now wrote to recount their own experiences of bad behavior. Military officers who had traveled with Marsh were also ready to back up his charges.

In mid-October, before the official investigating committee could release its report, Interior Secretary Delano resigned, and Smith soon followed. The report managed to avoid incriminating them, instead merely proposing to blackball certain contractors and to institute stricter contract and inspection procedures. The *Nation* explained that the investigating commission "was appointed by the very men whose official conduct was to be investigated, and the instructions issued to the court thus constituted were drawn up by the accused themselves." The *Boston Evening Transcript* noted, with both eyebrows raised, that the committee seemed to think it was a "fair and legitimate" profit for some lucky individuals to pocket $117,000 on the $700,000 contract for supplying third-rate beef to the reservation Indians.[13]

When Marsh carried Red Cloud's complaints to Washington, he made the government's mistreatment of Indians a national scandal. The two posed (with an interpreter at right) during Red Cloud's 1883 visit to New Haven.

It would be naïve to suggest that this affair produced dramatic long-term improvements for the American Indian. But Marsh had at least made it harder to ignore their appalling mistreatment. He emerged from the affair as a defender not just of the beleaguered Indians but also of the swindled American public. A craving for public adulation being one of his driving traits, he was undoubtedly pleased, and even more so when Red Cloud sent him a peace pipe soon after.

The accompanying note, dictated via translator to an army officer, said: "I remember the wise chief. He came here and I asked him to tell the Great Father something. He promised to do so, and I thought he would do like all white men, and forget me when he went away. But he did not. He told the Great Father everything just as he promised he would, and I think he is the best white man I ever saw. I like him. I want you to tell him this."[14]

In 1883, Red Cloud came to visit New Haven and renew his friendship with Marsh. The two posed for a studio photograph, and for the rest of his life Marsh prominently displayed a portrait of Red Cloud in his house.

CHAPTER 9

The Year of Enormous Dinosaurs

I wish you wer here to see the bones roll out and they are beauties.

W. H. REED to S. W. Williston, Como Bluff, 1878

On a Tuesday morning at the end of March 1877, a couple of amateur naturalists set out from Morrison, Colorado, to measure geological strata and collect plant fossils for a botanist back east. They were headed three miles up Bear Creek Canyon to a hogback ridge where "tier upon tier of yellow and brown and gray sandstones" rose up "like so many battlements defending the slope of the hill."[1] Arthur Lakes was an English-born teacher and itinerant minister who would become Colorado's foremost geologist and mining engineer. Henry Beckwith was a retired U.S. Navy engineer from Connecticut.

They had not climbed very high when Beckwith called Lakes over to see what appeared to be a fossilized piece of an ancient tree. Lakes, who had trained at Oxford, quickly saw that the surface was smoother than most fossilized trees, and "on one end were little patches of a purplish hue which I at once recognized as fragments of bone." The two men turned uphill to search for other evidence, Lakes wrote, "and as I jumped on top of the ledge there at my feet lay a monstrous vertebra carved as it were in bas relief on a flat slab of sandstone." It was thirty-five inches in circumference, "so monstrous . . . so utterly beyond anything I had ever read or conceived possible that I could hardly believe my eyes and called to my friend Captain B. to confirm the vision. We stood for a moment without speaking gazing in astonishment at this prodigy and threw our hats in the air and hurrahed: and then began to look about us for more."[2]

The year of enormous dinosaurs had begun.

Until then, almost nothing was known about dinosaurs or about the immense

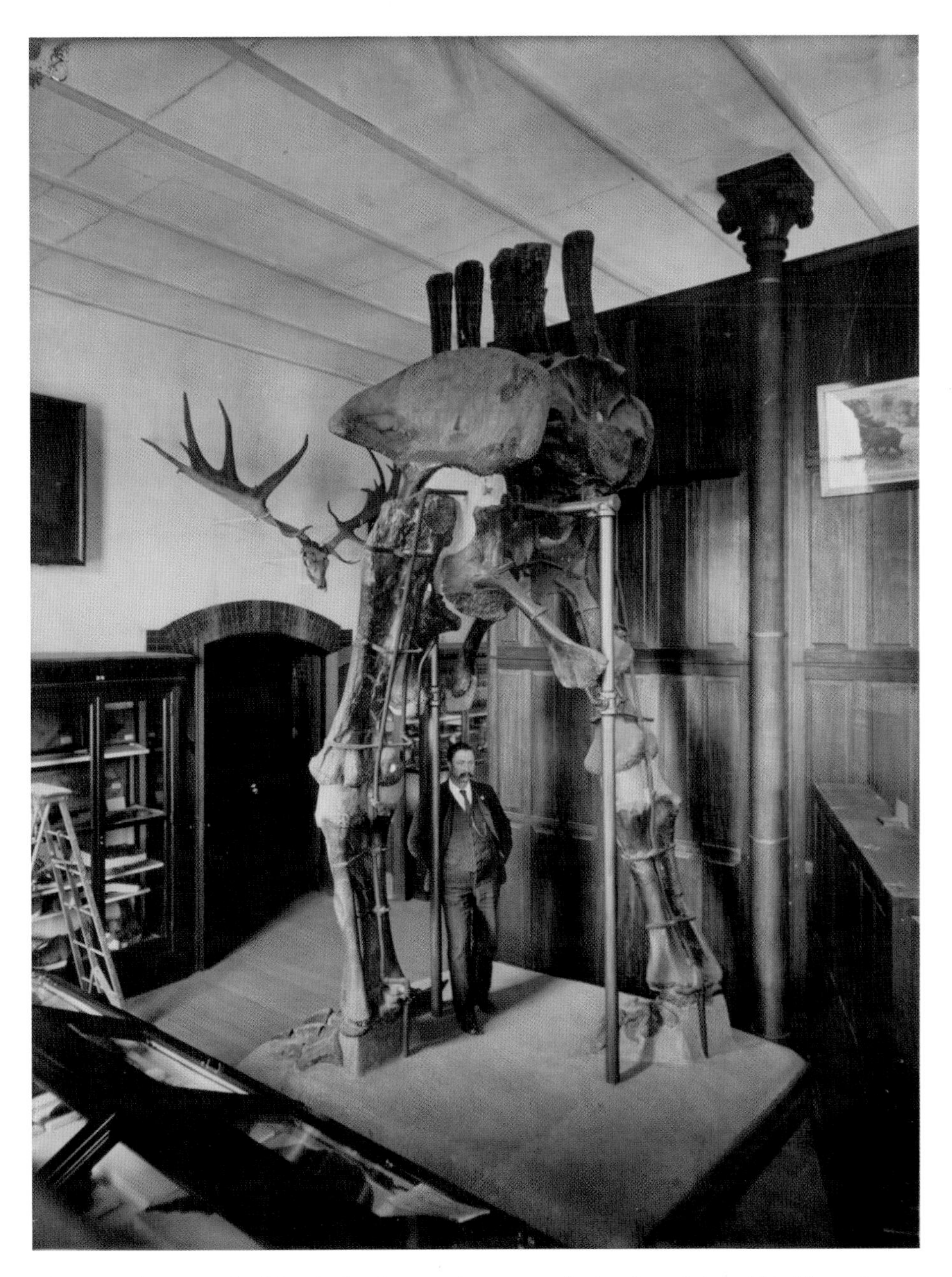

The original museum building had only enough room to display the pelvis and hind legs of *Brontosaurus* in a mount by preparator Hugh Gibb, who posed beside it for scale.

Arthur Lakes's watercolors give a firsthand look into a bone hunter's life, here showing a panorama of the Como Bluff landscape, with collector William H. Reed in the foreground.

scale some of these ancient creatures attained. The British theologian and paleontologist William Buckland had published the first scientific description of a dinosaur in 1824, and the word *dinosaur* itself, meaning "terrible lizard," dated only from 1842. The only dinosaur skeleton excavated intact, or nearly so, in North America was Leidy's *Hadrosaurus,* discovered in 1858. This patchy history made dinosaurs an oddity, with no set place in scientific thinking. Dana's *Manual of Geology* for 1865 covered the entire group in four brief paragraphs—while devoting ten pages to trilobites. The miraculous year of 1877, with monumental dinosaur finds at Morrison and Cañon City in Colorado and at Como Bluff in Wyoming, would change that forever, to the eternal delight of monster-loving children—and adults—everywhere.

The Morrison Formation

At the beginning of April, Lakes wrote to Marsh to report the discovery of "a vertebra and a humerus bone of some gigantic saurian in the upper Jurassic or lower Cretaceous." Marsh promptly replied offering to identify the specimen, but Lakes had "not yet quite decided what to do with the bones and I hope to add somewhat to their number before I do decide." Meanwhile, either unaware of the notorious rivalry or, perhaps, hoping to profit by it, Lakes sent samples of his discoveries to both Marsh and Cope. Alarmed by this development, Marsh sent off a $100 retainer in June and dispatched his collector Benjamin Mudge to preside over the dig and stake it out against interlopers. "Satisfactory arrangements made for two months," Mudge soon telegraphed Marsh. "Jones cannot interfere."[3] Jones, mean-

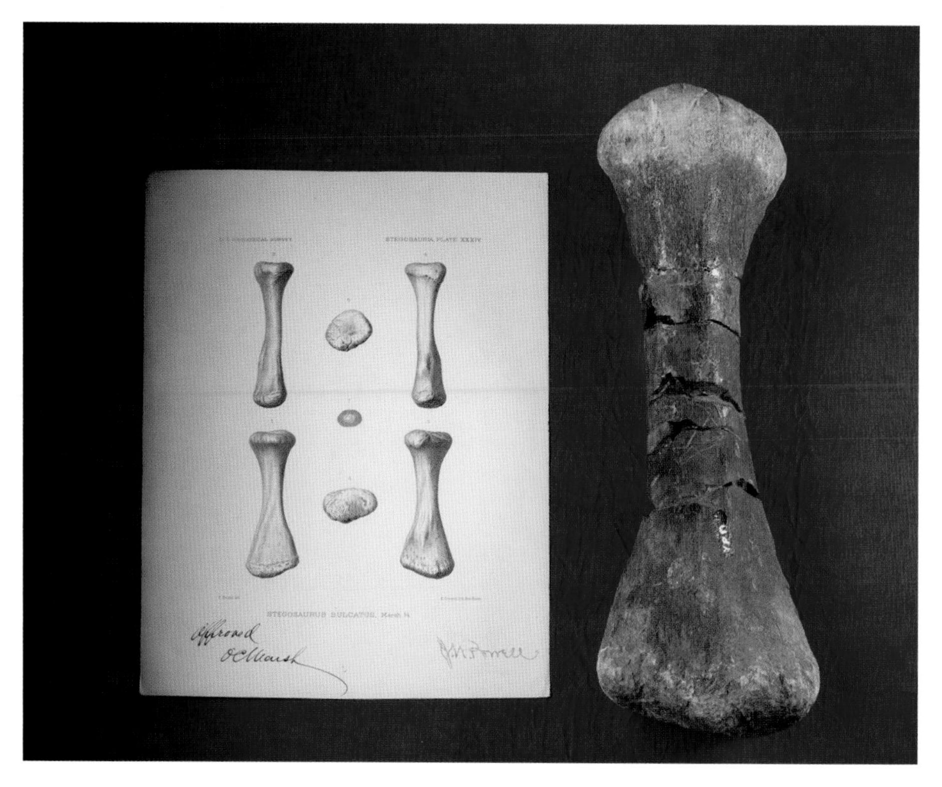

A radius from the left forelimb of a *Stegosaurus sulcatus,* depicted in a lithographic plate prepared under the supervision of O. C. Marsh.

ing Cope, had little choice but to do as Lakes now directed and send the proffered specimens on to Marsh.

On July 1, just three months after Lakes and Beckwith let out that first "Hurrah!" the *American Journal of Science* published Marsh's description of "a new and gigantic dinosaur," fifty or sixty feet long, which he named *Titanosaurus* (later *Atlantosaurus*). Specimens from Colorado also resulted later that year in Marsh's descriptions of the predator *Allosaurus* and the massive herbivore *Apatosaurus.* Morrison must have felt like the paleontological promised land. Among other finds was "one of the most remarkable animals yet discovered," in Marsh's words, an extinct reptile thirty feet long, with "large bony dermal plates" on its back. He called it *Stegosaurus.*[4]

The work was hard and at times dangerous. Lakes and his crew did much of their digging under an overhanging ledge that sheltered them from the Rocky Mountain winter as they burrowed deeper into the past. One evening, after snow had given way to a period of melting, the men left their tools in the dig and adjourned to the village below to eat and sleep. At about midnight, Lakes wrote, one of them "heard a roar like thunder from the cliffs above as if the whole hogback was coming down." When they arrived at the dig next morning, "lo there lay the whole ledge fallen in and filling the excavation with a weight of rocks over 100 tons." It took them a week to excavate the mass of stone—but it was better than being buried beneath it.[5]

Como Bluff

Marsh's previous work had focused mainly on ancient birds, marine reptiles, and mammals, not dinosaurs, and he was still dealing with the overwhelming abundance of strange new creatures from Colorado when a letter arrived in mid-July from southeastern Wyoming offering "a large number of fossils." The writers, who identified themselves as "Harlow & Edwards," were deliberately vague about where they had found the bones. "We are working men and are not able to present them as a gift," they explained, "and if we can sell the secret of the fossil bed, and procure work in excavating others we would like to do so." After seeing some eye-opening samples, Marsh sent a check for $75 and directed his collector Samuel W. Williston to the scene.[6]

The first thing Williston discovered was that "Harlow & Edwards" could not cash the check because they had been a tad too secretive: the bank knew them only by their real names—William Harlow Reed and William Edwards Carlin, Union Pacific Railroad workers at Como Station in Wyoming Territory. Williston also realized that they had stumbled onto a mother lode. "I have seen a lot of bones that they have ready to ship," he wrote to Marsh. "They tell me the bones extend for *seven* miles and are by the ton." Williston kept them at work into the winter, with a tent over the dig to protect them from snow, hoping to collect as many good bones as possible before word leaked out. "I wish you wer here to see the bones roll out," Reed wrote at one point when Williston was absent, "and they are beauties."[7]

From the nondescript little ridge called Como Bluff, two hundred miles north of the Morrison dig, Marsh would spin out a series of new dinosaurs, including *Laosaurus* and *Morosaurus,* in addition to collecting more complete and revealing specimens of species he had already described. Equally important, though less spectacular, was the tiny jawbone of a creature Marsh called *Dryolestes,* the first mammal ever found in North America from the late Jurassic period, about 150 million years ago. Altogether, the work at Como Bluff "probably had greater impact on the study of paleontology than any other event save publication of Darwin's theory," John H. Ostrom and John S. McIntosh, both affiliated with the Peabody, would later write. "The finds at Como Bluff revolutionized field and collecting procedures, generated a startling growth in paleontologic studies, stimulated great public interest, and left a permanent mark on the major museums of the world," which were soon vying to display enormous dinosaurs as a major public attraction.[8]

Cañon City

Cope, meanwhile, had latched onto his own dinosaur graveyard at Cañon City, Colorado, 120 miles south of the dig at Morrison and part of the same Jurassic formation. Another amateur naturalist, O. W. Lucas, had discovered it at almost the exact time Lakes and Beckwith were making their discovery in Como. For three such major dinosaur deposits to have turned up in the same year was hardly coincidence. The West was becoming settled, and the publicity generated by Marsh's and Cope's earlier discoveries had alerted newcomers to the possibility of scientific glory—and hard cash—to be gained by carefully searching the countryside. On the heels of Marsh's *Titanosaurus* paper that summer of 1877, Cope published his description of *Camarasaurus.* "This remarkable creature," he declared, "exceeds in its proportions any other land animal hitherto discovered," including Marsh's *Titanosaurus,* which Cope dismissed as not even a legitimate species.[9]

That same year Cope described a species of the dinosaur *Amphicoelias* based on a single 4.5-foot-high neural arch, the crosslike structure on a vertebra to which massive muscles had once attached. Cope estimated that the entire vertebra would have been about six feet high, but the fragment went missing and no other fossil evidence has ever been found. Confirming the size of this reported monster "may be nearly every sauropod paleontologist's Holy Grail," a 2014 study noted, but it was likely a case of "extreme over-estimation" and possibly "a typographical error" brought on by the rush for paleontological glory.[10]

The location of the three new dinosaur graveyards could hardly stay secret for long. Before heading to Como Bluff, Williston checked out Cope's find at Cañon City and collected bones there that would become the long-necked, whip-tailed monster Marsh named *Diplodocus.* Later that year at Como Bluff, he reported with regret "that Cope knows of this locality and the general nature of the fossils." The two rivals also knew of each other's technological innovations. That year, Williston abandoned the haphazard "pick, rake and sack technique," which tended to reduce everything to a jigsaw puzzle jumble, and devised a new way of holding fragile

Brontosaurus Redux

On July 26, 1879, geologist Arthur Lakes, working at Como Bluff, Wyoming, jotted a note in his journal: "Men came back with report of discovery of very big bones at a spot between quarries 8 and 9. Heavy thunderstorms hailstones fell the size of hens eggs. Telegraph wires broken."[a] Lakes and his fellow collectors, Bill Reed, Ed Ashley, and Ed Kennedy, soon sent O. C. Marsh what turned out to be a spectacularly large and, except for the head, nearly complete skeleton of a dinosaur.

Marsh saw that it resembled the huge, long-necked sauropod *Apatosaurus,* which he had described two years earlier. But the new specimen had fewer vertebrae in the sacral region, at the base of the spine, leading him to believe that it might be a new genus—and in his race for priority over rival Edward Drinker Cope, new species were good, new genera even better. Marsh called it *Brontosaurus,* and it became one of the most popular dinosaurs ever known.

Collectors Ed Kennedy and Bill Reed excavating *Brontosaurus* bones from quarry 10 at Como Bluff. Reed seems to have fired his shotgun into the brim of his own hat.

But in 1903, paleontologist Elmer Riggs at the Field Museum in Chicago took another look and argued that Marsh was mistaken: he had neglected to consider that the sacral vertebrae in tetrapod species, including humans, normally fuse as an individual matures. Marsh's two specimens were simply older and younger individuals of the same genus (though Riggs still regarded them as separate species). "*Brontosaurus* is a synonym of *Apatosaurus*," Riggs concluded, and according to the rules of scientific naming, that meant the later name *Brontosaurus* belonged in the wastebasket.[b] At least among scientists, Riggs's argument eventually prevailed. Marsh's own Peabody Museum shed some tears—and in 2003, after a hundred years, finally substituted *Apatosaurus* for *Brontosaurus* on the specimen that now stands as the centerpiece of the Great Hall of Dinosaurs.

But the public always resisted that change. For one thing, *Apatosaurus* ("deceptive lizard") is a highly forgettable name. On the other hand, *Brontosaurus* ("thunder lizard") has what modern marketing types would call stickiness, sometimes literally: in 1989, the U.S. Postal Service issued a stamp with an image of the dinosaur labeled *Brontosaurus*, defending its name choice as "more familiar to the general population."[c]

The U.S. Postal Service, the general public, and O. C. Marsh may, it now appears, have been right all along. After five years spent carefully measuring and comparing different *Brontosaurus*, *Apatosaurus*, and other diplodocid dinosaur specimens at museums around the world, a team of European scientists announced in 2015 their conclusion: *Brontosaurus* is different enough to be a separate genus from *Apatosaurus*, and deserves its own branch on the dinosaur tree of life. They are as different, Peabody paleontologist Jacques Gauthier explained, as a robin and a bluebird.[d]

When the new analysis appeared in the journal *PeerJ*, one of the authors remarked that it had taken a while for people to accept that *Brontosaurus* was a taxonomic has-been, and "I guess it will also take some time" for people to accept that it's back.[e] But just one week later, the Peabody Museum was ready with what it called a "Re-renaming Ceremony." As he drew back a veil to reveal a sign restoring the dinosaur's original name, Gauthier called out, "*Brontosaurus* is back!" to cheering from the crowd. Then he added that the real restoration was still to come, when preparators remount the *Brontosaurus* specimen in a more dynamic (and realistic) pose, minus the cumbersome steel understructure of the early 1930s. That process, he noted, will cost $1.8 million, "and I take checks."[f]

dinosaur remains together. "Will it do to paste strips of strong paper on fractured bones before removing?" he wrote to Marsh, adding, "These strips are put on with ordinary flour paste and can be removed I think easily."[11] But Cope's side may have gotten there first when his collector Charles H. Sternberg patched bones together with burlap and rice paste.

The two sides now often worked in close proximity, and both engaged in wildly unprofessional behavior. They spied on each other, of course. They obliterated their adversaries' place markers. They planted false clues to sow confusion. They even scattered bones from an assortment of species close together in the hope that they might be concocted into a bogus new species. (Cope fell for it and published a species description, later retracting it after a younger colleague gently pointed out the deception.) They destroyed fossils, at least once on indirect orders from Marsh, to prevent the other side from finding what they could not collect themselves. And of course they hired away each other's workers. William Carlin, who still had his job as stationmaster at Como, defected from Marsh to Cope. Thereafter, reenacting the rivalry of their employers, Carlin would carefully pack specimens for Philadelphia in the comfort of the station, while Reed had to do his packing for New Haven in the frigid Wyoming winter out on the open platform, an experience he described, with remarkable restraint, as "meerley H——, H——, H——."[12]

In the circumstances, strong language was not unknown to these gentlemen. Cope, who was still a Quaker, felt obliged to forbid swearing in camp; at times he also read the Bible aloud to his crew. Marsh merely "made a solemn vow never to use any profanity east of the Missouri River," but he could not always help himself. One time at the Peabody, he put down some new paper by Cope and exclaimed, "*Gad!*" Pause. "*Gad!*" Pause. Then, explosively, "*Gad! Goddammit! I wish the Lord would take him.*"[13]

In July 1878, under the eye-catching headline "Gigantic Reptiles and Dragons," the *New York Times* ran a story about the past year's sensational finds, which were proving to be "of much greater importance to the scientific world than many who are well versed in geology had supposed." A reporter for the *Omaha Bee* had tracked down Williston at his hotel and extracted an interview. "The remarkable thing about our discovery," the bone hunter told the reporter, was that fossils previously thought not to exist in America "had hardly been brought to light in one locality before thousands of tons of remains were simultaneously discovered" in a half dozen different places. With some of Marsh's flair for publicity, Williston described one such specimen of "what is termed the dinosaur" standing as tall as Omaha's five-story Grand Central Hotel, while another "must have been about

100 feet in length." A "flying reptile, or dragon" had wings "30 to 40 feet from tip to tip." To the reporter's question about the cost of all the collecting, Williston answered that Yale had already spent $250,000. He did not say that much of it had come out of Marsh's own pocket, in the ten years since his first western foray to that well at Antelope Station.[14]

Rancor in the Workroom

The decade that followed was a time of triumph for Marsh and precipitous decline for Cope. Marsh had always understood the power of social and political connections and the importance of exciting the attention of the general public. Cope, not unreasonably, thought that solid science should be enough, and he shrank from presenting his discoveries in terms that might have helped people grasp their considerable importance. His difficult, quarrelsome personality also won him few friends in politics or the academic world.

In April 1878, Marsh became vice president of the National Academy of Sciences, having been elected to membership just four years earlier—over Cope's strenuous objections. Then as now, the academy wielded great influence over all aspects of science, and Marsh was soon working through it to consolidate government surveys and put them in the hands of scientists rather than the military. Cope was the lone vote against this plan, arguing that the geological survey should be left as it was. In 1879, after considerable lobbying by Marsh, his Yale contemporary Clarence King became head of the U.S. Geological Survey, weakening Cope's already minimal influence as a government paleontologist. King promptly published Marsh's 201-page monograph *Odontornithes,* detailing the anatomy of the toothed birds. The lavish illustrations commissioned by Marsh depicted, among other things, the *Ichthyornis* specimen Mudge had originally addressed to Cope—then shipped to Marsh instead. For Cope, there was worse to come: when John Wesley Powell, another Marsh ally, succeeded King in 1881, Marsh became the survey's new vertebrate paleontologist, with access to roughly $10,000 a year in federal funds for collecting. Marsh soon rose to the presidency of the National Academy, a position he held from 1883 to 1895. His prep school "superiority in managing practical affairs" was bearing plentiful fruit.

About this time, Cope began to encounter financial difficulties. He had inherited $250,000, a fortune, on his father's death in 1875. But just four years later, he was in need of money thanks to the pell-mell pace of discovery and his abiding urge to get ahead of Marsh. Made overconfident by his knowledge of geology,

Cope began to invest in the fool's gold of western mining companies. When these investments inevitably failed to pan out, he was left with no income to support his wife and daughter in the Copes' houseful of fossils on Pine Street in Philadelphia.

Attending the annual meetings of the National Academy, chaired by Marsh, must have been agony for him. Marsh presided with "a dignified formality," according to Russell H. Chittenden, a physiological chemist at Yale. Though he might appear "somewhat stern," "rather punctilious," and with "a stiffness of bearing" that others frequently misunderstood, "Marsh nevertheless possessed an innate courtesy and kindness of heart which . . . made him a friend and colleague to be respected and admired." Another scientist put it in less charitable terms: "Many regarded him as pompous and vain. His relations with his colleagues were not always friendly."[15]

This last point would turn out to be a considerable failing. For all his skill at cultivating the friendship of "those whom it was most desirable to know," Marsh treated his Peabody Museum subordinates as mere functionaries, though among them were such scholarly figures as Williston, George Baur, Oscar Harger, Erwin H. Barbour, and later John Bell Hatcher. "Marsh's method of work," according to his biographers Schuchert and LeVene, "was to assign to each assistant a certain series of fossil bones," typically with instructions "to have the nearest affinities of these animals sought for, either in the literature" or in his own fossil and bone collections. "In due time he would sit down with each man, asking questions, discussing conclusions arrived at, and all the while making notes in regard to what he heard and what he saw in the assembled material." Many of these assistants were as knowledgeable and "as ambitious as he to publish research. The dissatisfaction arising therefrom was aggravated by his habit of keeping his own conclusions strictly to himself in these conferences, especially when he differed with his staff; he rarely exposed his final conclusions until they appeared in print."[16]

Marsh not only frustrated his colleagues' desire to publish at least some of their work under their own names, he also paid them poorly and with an irregularity, a Marshiness, that kept them constantly dependent and on edge. Instead of understanding that the success of the people he trained was an essential part of any scientist's enduring legacy, he did everything possible to delay or prevent their advancement in the academic world. Williston later recalled that, by 1878, "I had given loyal service to Professor Marsh for four years, had given up my medical career, with no prospects before me but those of a field collector, with a present salary of $40 a month and a possible one years hence of $100, with an opportunity of doing all the research I desired to be published by him!"[17] And yet he remained at the Peabody until 1885, no doubt in part because of the excitement of the great discoveries being made there. The atmosphere in the workrooms was rancorous as a result, and this festering mood became one more secret that could not be kept.

Marsh also managed to antagonize a younger generation of scientists from outside the Peabody, notably William B. Scott at Princeton and Henry Fairfield Osborn at the American Museum of Natural History, both friends of Cope. Though he did not detail the circumstances, Scott had a run-in with Marsh's "egoism, his extreme selfishness and unscrupulous duplicity" in 1877, which even decades later left him "closer to hating him than any other human being that I have known." Marsh used his U.S. Geological Survey position in an effort to block Osborn, and probably Scott, too, from exploring public and Indian lands. The rage Marsh instilled in these two scientists would come back in time to haunt him. In this developing drama, Marsh was the tragic hero blinded by arrogance and an insatiable ego. Cope, his relentless adversary, was "a character out of fiction, a distinguished scientist with an emotional life like that of the villain of a Jacobean tragedy," or so Wallace Stegner wrote. Scott and Osborn would play the part of Cope's loyal henchmen, whispering behind the arras, plotting and waiting to drive the knife home.[18]

Cope had been obsessed with besting Marsh for years, but in the 1880s his thoughts bent toward revenge, and he began to seek out and file away the ugly rumors he was hearing from disgruntled younger scientists at the Peabody. During those years as an outcast, he and Osborn sometimes met in Philadelphia for lunch at a local spot where they could eat for 25¢, following their meal with a cigar. Osborn had no money worries—his father was a railroad tycoon—but he noted that Cope always returned the cigar to its case half smoked, "for future reference." He added that "small economies of this kind were shown in the case of paper, and twine, in fact everything." In Cope's crowded study, Osborn observed how "the accumulating fossils, books, and pamphlets outtaxed the shelves and began to thicken like stratified deposits upon the floor in dust-laden walls and lanes."[19]

Marsh's name of course came up over and over in conversation, and Cope "showed far more humor than bitterness" about his great rival, Osborn recalled, though the letters Cope was writing at the time suggest he felt otherwise. Then "one day he slyly opened the lower right hand drawer of his study table and said to me: 'Osborn, here is my accumulated store of Marshiana. In these papers I have a full record of Marsh's errors from the very beginning, which at some future time I may be tempted to publish.'"[20]

CHAPTER 10

Fossils, Buffalo, and the Birth of American Conservation

Their white skulls dot the prairie in all directions.

GEORGE BIRD GRINNELL, *Zoological Report,* 1874

The conventional view of the American conservation movement is that it started in the late nineteenth century with hunters alarmed by the rapid disappearance of game animals. But it also started with fossils, or rather, with George Bird Grinnell and what he called "fossilizing" in the American West. On his first trip there with O. C. Marsh in 1870, and on return visits for the Peabody Museum, Grinnell saw at least three things that would profoundly influence the rise of the conservation movement in the United States. All three, curiously, were about undoing the worldview of the naïve young man who set out on that expedition, as Grinnell himself put it, thinking that fossils "meant nothing to us," and wanting only "to shoot buffalo and to fight Indians."[1]

Grinnell was born in New York City to a blueblood New England family, the son of a stockbroker whose firm was the chief trading representative for Cornelius Vanderbilt. His fascination with the natural world began in the Upper Manhattan neighborhood where he grew up, Audubon Park, between modern-day 155th and 158th streets. With earnings from his *Birds of America,* John James Audubon had purchased land and established his family there in 1841, when the area was still woods and farmland. His widow, Lucy, then sold off parcels to wealthy families like the Grinnells as it developed into a leafy exurb. Young Grinnell spent much of his free time playing in the Audubon houses and barns amid the residue of the artist's career. He came to know the younger Audubon son, John Woodhouse Audubon, "a bluff, gruff, but friendly man" who "was always willing to talk about birds,

George Bird Grinnell accompanied Marsh on his 1870 fossil expedition and went on to help found the American conservation movement.

mammals, or, indeed, any natural history object, to any boy who asked him questions."[2] Lucy Audubon schooled local children in her home (they called her Grandma Audubon), and she instilled an ethos of self-denial that Grinnell would later echo in his writings about hunting and conservation.

When his more formal schooling began, Grinnell proved an indifferent student. His father pressed him to go to Yale even so, and with the help of expert tutoring and a lot of last-minute cramming, Grinnell squeaked through the entrance exam. After that, he skittered along on the scholastic fringes. He managed to avoid punishment for an incident in which he painted his class year on the face of the campus clock, but Yale rusticated him for a time to an off-campus study site for his involvement in a hazing incident. He was barely able to escape in 1870 with his college degree, in a dead heat for second to last in his class.[3]

His subsequent expedition with Marsh changed his life. It was of course the hunting that hooked him at first, along with the experience of open country and the exposure to western characters like the ones in the adventure novels he had read as a boy. But something also seemed to shift in him as he dug out the fossil remains of extinct crocodiles, turtles, and early mammals. When the expedition ended, Grinnell dutifully joined his father at the family's brokerage, despite "a settled dislike for the business."[4] By 1874, though, he was back at Yale as a graduate student and unpaid assistant to Marsh. He was thus a witness as the crates coming in from the West opened to reveal entire lost worlds of astonishing creatures. There was hardly a better place on Earth to inculcate an abiding awareness that almost any species could become extinct.

"Grinnell was in the vanguard of the ascending generation of more sophisticated explorers," the historian and Grinnell scholar John F. Reiger wrote, "and his awareness of the great transformations the earth had undergone made him recognize that the land was not invulnerable as most of his contemporaries seemed to believe. His study of fossil animals taught him that the long-term survival of a species was the exception, not the rule. When the concept of a vulnerable natural world was taken out of the past and applied to his own time, it naturally led to a new concern for the land."[5]

"These Shaggy Brown Beasts"

One of the great revelations for Grinnell on the first leg of that 1870 expedition was the experience of traveling with the U.S. Army patrol's two Pawnee guides and coming to know them as hunting companions and friends. He went west again to hunt with the Pawnees in the summer of 1872, and in northern Kansas, "among the numberless bluffs that rise one after another like the waves of a tossing sea," he saw buffalo by the thousands, "some peacefully reposing on the rich bottoms, others feeding upon the short nutritious grass that clothes the hillsides." He also rode along on a traditional buffalo hunt with a party of eight hundred Pawnee braves, men who could ride amid charging buffalo and fire arrows repeatedly from horseback, sometimes powerfully enough for the arrow to pass through a buffalo and protrude from the opposite flank. The hunt, as Grinnell recalled it, was a potent display of tactical skill and discipline, geared to take only as many animals as needed.[6] It would also turn out to be that tribe's last great buffalo hunt. By the end of 1873, newcomers with a very different attitude had eradicated the buffalo from Kansas. Though it seemed impossible, they would soon do so throughout the Great Plains.

Grinnell had begun to write for a new weekly magazine called *Forest and Stream,* and in its pages he would in time become one of the nation's leading voices for protection of the buffalo. (William Hornaday of the Smithsonian Institution and later the New York Zoological Society was the other great advocate for the buffalo.) A recurring theme for Grinnell was the contrast between the appallingly wasteful slaughter of buffalo then being perpetrated by settlers and the Indian hunter's economical practice of using almost every scrap of what he killed. He may have exaggerated the conservationist ethic of Native Americans or emphasized only those parts that fit his own pragmatic worldview. Regardless, Grinnell made the Indian hunter an important model for reforming the behavior of supposedly civilized hunters and for stopping the senseless killing of the buffalo: "Unless some action on this subject is speedily taken not only by the States and Territories, but by the National Government," he warned in an 1873 article, "these

In 1873, George Grinnell witnessed the skill and restraint of an Indian buffalo hunt in Kansas (much as George Catlin had once painted it). But it was the last hurrah for such hunts.

shaggy brown beasts, these cattle upon a thousand hills, will ere long be among the things of the past."[7]

In 1874, Marsh sent Grinnell west again as his substitute to join the expedition into the Black Hills with George Armstrong Custer's Seventh Cavalry. Custer took a liking to the young scientist and sometimes invited him to dine. At one point, he sent an enthusiastic dispatch remarking on Grinnell's discovery of "an important fossil . . . a bone about four feet long and twelve inches in diameter," which "had evidently belonged to an animal larger than an elephant." (No record of this fossil has survived.) Grinnell in turn spoke of Custer's "indomitable energy and perseverance," but otherwise did not return this admiration. When someone mentioned during a campfire conversation how lucky they were not to have encountered a large party of hostile Indians, Custer boasted, "I could whip all the Indians in the northwest with the Seventh cavalry," a claim that would soon be put to the test. Grinnell predicted more accurately that the expedition and its discovery of gold in the Black Hills would lead to "an Indian war I expect, for the Sioux and Cheyennes won't give up this country without a fight."[8]

For Grinnell, the more immediate problem was that Custer often kept the expedition on the move from before dawn to well past dark. "It is almost useless to attempt to collect fossils," Grinnell griped in his diary for July 16. "The main object

. . . is to reach the Black Hills, and the commanding officer cannot wait for fossils." On July 17, Grinnell drily noted, "Race day at Saratoga started about 5 a.m." Finally, on August 23, his last collecting day on a trip that would cover an average of almost fifteen miles a day for sixty days, he wrote: "Thus ends my expedition of 1874 as far as science is concerned: a total failure." The expedition had not seen a single living buffalo in a country that had been a "favorite feeding-ground" only a few years before. Now "their white skulls dot the prairie in all directions." In one place, Indians had painted them and arranged them ritually "in five parallel rows of twelve each, all the skulls facing the east."[9]

Despite the 1874 "failure," Marsh "seemed to think it desirable" for Grinnell to join an expedition into Montana and Wyoming the following summer with the Seventh Cavalry's Colonel William Ludlow. Grinnell's friend and fellow graduate student Edward S. Dana, son of Professor James Dwight Dana, signed on as geologist. Predictably, the pace of the expedition, on the same sixty-day schedule, once again proved too fast for proper fossil hunting. Ludlow felt compelled in his report "to express regret at Mr. Grinnell's comparative disappointment in not securing a larger collection of fossils."[10] But this time the trip was grimly worth it.

Yellowstone National Park, the ultimate destination for this expedition, had been designated in 1872 as the world's first national park. But it was also the first "paper park," little more than lines on a map, with no real rules and no one to enforce them. The expedition arrived in August to find whiskey dealers and poachers camped out in the middle of the park, engaging in open "butchery" of the ostensibly protected wildlife. "Buffalo, elk, mule-deer, and antelope are being slaughtered by thousands each year, without regard to age or sex, and at all seasons," Grinnell wrote in the "Letter of Transmittal" accompanying his official report. "Of the vast majority of animals killed, the hide only is taken," to be sold for $2.50 or $3 apiece, with the rest of the carcass left to rot or be scavenged by wolves. "Females of all these species are as eagerly pursued in the spring, when just about to bring forth their young, as at any other time." This scene of "wholesale and short-sighted slaughter" would haunt Grinnell for the rest of his life, driving him to campaign relentlessly for some of the most important wildlife protection laws in American history.[11]

The following May, General Custer sent an invitation by telegram for Grinnell to join another military expedition, this time up the Yellowstone River to the Big Horn Mountains. After consulting with Marsh, Grinnell telegraphed back that he was simply too busy with work at the Peabody. It was one of only three summers in the 1870s that he passed up a chance to go west, and his regret at doing so was genuine. As a result, Grinnell was not present on June 25, 1876, when the friends he had made in the Seventh Cavalry, "always jolly and friendly and so thoughtless of the future," lost their lives with Custer in the battle of the Little Big Horn.[12] Grinnell's good friend "Lonesome" Charley Reynolds, a civilian scout, was among the first to die.

Saving the Buffalo

That same year, Grinnell agreed to become the natural history editor of *Forest and Stream* while continuing his work with Marsh at the Peabody Museum. He was to provide a page or two in the magazine every week and at the same time help produce Marsh's first great work, *Odontornithes: A Monograph on the Extinct Toothed Birds of North America,* published by the U.S. Geological Survey in 1880. No doubt to the astonishment of his former instructors, Grinnell was then also successfully pursuing his doctorate, with a thesis, "The Osteology of *Geococcyx Californianus,*" about the greater roadrunner. It was a heavy workload, and Grinnell began to suffer sleeplessness and headaches. A "nerve and brain specialist" in New York warned him to change his line of work, or "be prepared to move into a lunatic asylum or the grave." Grinnell and his father had meanwhile been buying up shares of Forest and Stream Publishing Company stock, to the point that they now held a controlling interest. When the opportunity arose in 1880 to become president and editor in chief, Grinnell "left New Haven with keen regret, for I had hoped to work for a long time in the [Peabody] Museum."[13]

His true calling lay elsewhere, campaigning against the destruction of the American West, which he foresaw as if in a vision: mines would ravage the landscape, he wrote in an 1879 article, with "a thousand smelting furnaces" to blacken "the pure, thin air of the mountains." The game animals, "once so plentiful, will have disappeared with the Indian." Railroads would climb the hillsides bringing supplies to the mining towns and return "freighted with ore just dug from the bowels of the earth . . . all arable land will be taken up and cultivated, and finally the mountains will be stripped of their timbers and will become simply bald and rocky hills."[14]

For the next twenty years, he worked to forestall this fate, focusing first on laws to protect Yellowstone National Park. He fought down a bid by a group calling itself the Yellowstone Park Improvement Company to grab a one-square-mile monopoly around each of the park's major attractions including an unlimited right to cut timber in the park for an annual rent of $2 an acre. In addition to writing numerous editorials, he traveled to Washington, D.C., to lobby congressmen—

"the meanest work I ever did." In 1893, after two decades of "struggle and sweat and fight," Grinnell and his allies defeated a plan to run a railroad through the park.[15] The following winter, Grinnell sent a reporter to Yellowstone to write about the status of the plains buffalo, by then eradicated from all of North America except for a small herd in the park's Pelican Valley. Scarcity only made them a more profitable target for poachers.

Funds to protect the buffalo and other wildlife within the park were nonexistent. But shortly before Grinnell's reporter arrived, the park superintendent paid out of his own pocket to send out a two-man antipoaching patrol on skis. In mid-March, the patrol crossed the tracks of the park's most notorious poacher, Edgar Howell of Cooke City, and followed them into Pelican Valley. There, they spotted Howell skinning one of five buffalo he had just killed, among the last of the species. The lead scout, Felix Burgess, managed to ski across hundreds of yards of open country, at one point leaping a ten-foot ditch, to arrive undetected between Howell and his rifle. "I called to him to throw up his hands," Burgess told Grinnell's reporter, with obvious satisfaction, "and that was the first he knew of any one but him being anywhere in that country. He kind of stopped and stood stupid like, and I told him to drop his knife." Howell's arrest was "undoubtedly one of the best things that ever happened for the park," according to Grinnell, who heard about it almost immediately by telegram from his reporter.[16]

Next day, Grinnell headed to Washington, D.C., to undertake more of "the meanest work I ever did," while also ensuring that *Forest and Stream* and other publications "thoroughly exploited" the arrest, elevating Howell's actions to the status of a national scandal. Grinnell had cultivated powerful allies over the years, among them a New York assemblyman named Theodore Roosevelt. Together, he and Roosevelt had cofounded the Boone and Crockett Club, an organization of wealthy and well-connected eastern big-game hunters dedicated not just to "manly sport with the rifle" but to the passage and enforcement of laws protecting wildlife. Now these forces came together around the man Roosevelt described as that "infernal scoundrel," Edgar Howell.[17]

The park superintendent then had no power to do much more than escort Howell out of the park and admonish him not to come back. But within a week of Howell's arrest, John F. Lacey, a Republican congressman from Iowa, introduced a bill to outlaw the taking of wildlife in Yellowstone National Park and to authorize a jail term of up to two years for each offense. Within two months of Howell's arrest, that bill became law. "The year 1872 marked the birth of Yellowstone as a place," one historian wrote, "but 1894 marked the birth of Yellowstone as a commitment."[18]

That law was also the salvation of an American icon. The park's plains buffalo—the last remnant of the 30 million or so that had roamed across North America at the start of the nineteenth century—continued to decline for a few more years, to a low of just twenty-three individuals at the start of the twentieth century, about as close to extinction as a species can come without passing over into oblivion. But from there the plains buffalo has slowly recovered, to a population today of forty-nine hundred purebred buffalo within Yellowstone itself, perhaps thirty thousand in North America, and another five hundred thousand that are to some degree crossbred with domestic cattle.

Feathers and Beyond

Grinnell had meanwhile been campaigning on many other fronts. In 1883, his *Forest and Stream* began "hammering away" at the fashionable practice of ornamenting women's hats with feathers—and even whole birds and nests. When the public began to see—but too slowly—that this practice was "abominable," he founded the first Audubon Society, dedicated to protecting wild birds and discouraging "the wearing of feathers as ornaments or trimming for dress." Grinnell thought legislation could do little to stop "this barbarous practice," but he pushed for a change in public opinion, to be "inaugurated by women." He was no doubt patronizing in his assurance that "their tender hearts will be quick to respond" to the slaughter of birds. But the women he hired to write for his new magazine, also called *Audubon,* came to the task with talons extended. The poet Celia Thaxter described one unpersuadable woman going on her way with "a charnel house of beaks and claws and bones and feathers and glass eyes upon her fatuous head." Grinnell's Audubon Society and its magazine both eventually went out of business. But soon after, new Audubon societies began to spring up at the state level, renewing the fight, often with strong leadership by women. By reaching out to women, Grinnell "played a central role," historian Carolyn Merchant wrote, "in bridging the gender gap" in the conservation movement.[19]

In 1894, Grinnell's *Forest and Stream* moved beyond the hat fight with a broad proposal to ban all commercialized killing of wildlife: "Why should we not adopt as a plank in the sportsman's platform," an editorial asked, "a declaration to this end—*That the sale of game should be forbidden at all seasons?*"[20] This effort ultimately encouraged Representative John F. Lacey to push through Congress a broad new law to limit trafficking in plants and animals nationwide. More than a century later, the Lacey Act remains the nation's fundamental law for the protec-

The *Shortgrass Plains* diorama at the museum features three buffalo on the Wyoming ranch of William "Buffalo Bill" Cody.

tion of wildlife. Grinnell was also a major influence in promoting the Migratory Bird Treaty with Great Britain in 1916, which currently protects about eight hundred bird species, and on the passage of conservation-based hunting laws in many states.

Some recent historians have criticized Grinnell and his Boone and Crockett allies as a wealthy eastern elite imposing their will on distant landscapes at the expense of local people, who were often working-class or subsistence hunters. The charge seems true enough, and according to historian Adam Rome, the language in *Forest and Stream* was often tainted with class and ethnic overtones. The term *pothunter,* for instance, became a euphemism for immigrant hunter, and "the most reviled were the Italians."[21] But it seems doubtful that anyone today (including this author, whose great-grandfather was one of those Italian pothunters) would choose the alternative, which was Edgar Howell at his bloody work in Pelican Valley. The alternative would have been to allow the buffalo to precede the passenger pigeon in the unbelievably sudden rush from ubiquity to extinction. The alternative was to forego the creation of Glacier National Park and the Adirondack Forest Preserve, both also largely Grinnell's doing.

Moreover, in books about the Pawnee, the Cheyenne, and the Blackfoot tribes, "Grinnell, as much as any man of his generation, worked diligently to commit Indian memories to the page and thus preserve them," historian Sherry Lynn

Smith wrote in her 2000 book *Reimagining Indians.* "He attempted to accomplish this with minimal interference and embellishment from 'outsiders,' including himself. . . . Trained as a scientist, not a moralist, Grinnell wanted to log Blackfeet's, Pawnees,' and Cheyennes' words, actions, practices, history, and religious beliefs as objectively, accurately, and faithfully as possible." His training had made him, in short, an anthropologist. As a result, Smith continued, "Grinnell's writings helped lay the foundation for a fundamental reassessment of Indians and thus contributed to an evolving perception of them as people worthy of interest and respect. . . . Grinnell's essential message was this: Plains Indians were human beings with histories of their own, full lives, and complete cultures. If no one could stem the inevitable tide of evolutionary change ushered in by conquest, at least readers should consider and appreciate what had been destroyed."[22]

Grinnell's accomplishments in maturity represent an extraordinary transformation from the fecklessness of his privileged youth. And maybe that sea change would have happened without fossils, without the expedition of 1870, without O. C. Marsh. But it seems more plausible to suggest that George Bird Grinnell's career vindicated both Lucy Audubon's work as a teacher of young children and George Peabody's investment in science and education at the Peabody Museum. At second hand, they had helped to remake the world.

CHAPTER 11

A Building of Their Own

We feel like the crab just ready to break loose from its old shell—and enjoy the freedom of space adapted to a new and higher condition.

JAMES DWIGHT DANA to O. C. Marsh, July 27, 1865

The need for the museum building George Peabody had endowed became year by year more urgent. Various scientific collections were scattered around the Yale campus, some in the old Cabinet Building, some at the Sheffield Scientific School, others in basements or inaccessible storage spaces around the campus. They were also growing. In 1869, the trustees of the (still imaginary) Peabody Museum reported the arrival of fifty thousand new specimens over the past year, and this was before several new faculty members—notably O. C. Marsh in paleontology and Addison E. Verrill in invertebrates—had begun to collect in earnest.

Peabody had, however, allotted only $100,000 of his gift for construction, with another $30,000 set aside for maintenance and $20,000 to accumulate interest as a reserve building fund for a later addition. This amount was "not yet sufficient to provide such a building as the university requires," the trustees concluded in 1869, opting to delay construction a little longer.[1] Peabody had made his gift with his usual prudence, in the form of 5 percent bonds, payable in gold. Despite the Panic of 1873 and the ensuing depression, the bonds continued to appreciate reassuringly, and the trustees finally moved to break ground in 1874. (By the time the new building was completed in 1876, Peabody's original $100,000 had grown to $178,524.)

In a mark of the increasing stature of science, the university offered the Peabody a prime location on the corner of the central Old Campus quadrangle, where College and Chapel streets meet, just across from the New Haven Green. It was

The Peabody trustees commissioned a museum design suited to their grand mission of understanding life on Earth.

the most conspicuous site on campus and the social center—here upperclassmen gathered along the old Yale fence, and here visitors first arrived when coming from the train station. The intellectual legacy of Benjamin Silliman Sr., carried forward by the formidable duo of Marsh and Dana, seemed about to reach its fruition. But the trustees—including Marsh, Dana, and the mineralogist George J. Brush—soon realized that this site, with the street on one side and the open quadrangle on the other, would require frontage detailing on all sides, resulting in considerable expense. The location also precluded the possibility of a service entrance in back big enough for—but they could hardly have imagined how big it would need to be. The trustees opted instead for a more practical location at the opposite corner of the Old Campus and just across the street.

Marsh still had grand architectural ambitions, and the final design for the museum, by J. Cleaveland Cady of New York, was an imposing structure in the Gothic style, with two wings extending out from an ornate central building, above which four steeply sided spires rose skyward. There was, however, only enough money at first to build the north wing—brick, with Nova Scotia sandstone detailing—running 115 feet along High Street and 100 feet deep down Elm Street.

The scientists now had thirty-four thousand square feet of space in which to pull together the various collections and display at least a fraction of them advantageously. If Marsh still aspired to the scale of Berlin's Museum für Naturkunde, Dana was more modestly concerned with the proper presentation of specimens in the space they now possessed. He mapped out the interior arrangement in detail,

But the available funds were enough to build only one wing of the original design.

scratching a note on the side of one page "that a series of apartments is better for showing off collections than one large hall. The architecture of the great hall may be grand, but the specimens of natural history arranged in it will look diminutive." Brick walls and hanging iron doors, framed in wrought iron, divided these rooms. Tiles meant to be fireproof covered wooden structures. A building "made to last at least 1000 years" was still the ambition.[2]

Skeletons and Skulls

The layout made no strict separation between scholarly and public space, as would later become standard for almost all museums. The Peabody was a teaching museum, with the fossil dinosaur footprints from the Connecticut River Valley displayed amid the working rooms in the high-ceilinged basement, and the minerals (in twenty-five cabinets plus cupboards and drawers) beside the lecture room and laboratories on the first floor. Geology and paleontology occupied the second floor and zoology the third, with anthropology and archaeology relegated to the attic.[3]

The curators had taken great care, *Scientific American* reported soon after the new museum opened in July 1877, "to put the exhibition on easy terms with the

The lecture room of the first Peabody Museum displayed Marsh's "horse pedigree" on the left wall and a reconstruction of the toothed bird *Hesperornis* at the front.

public," so that a reasonably informed visitor "will be instructed as well as gratified by even a hasty survey." Everything was organized systematically. The minerals followed Dana's scheme of classification by chemical properties, and the geological displays progressed stratigraphically. In zoology, Darwinian evolution served as the guiding principle, without apology or obfuscation, to a degree that seemed to unsettle the *Scientific American* reporter: "And now—how shall I state it tenderly, without giving offence?—the Primates of science, unlike those of the Church, include, with man, the monkeys." Looking down the row of skeletons and skulls, he added, "it is not easy to say at a glance where man ends and ape begins."[4] *Ichthyornis* and *Hesperornis,* recognized by the reporter as "the connecting links between reptiles and birds," figured prominently in the displays, along with the pterosaurs. The evolution of the horse was also carefully delineated, culminating in the skeleton of a champion Arabian racehorse named Esnea.[5]

But the freestanding dinosaur skeletons for which the Peabody would later become renowned were absent then. Marsh's most spectacular dinosaur finds were still a year or two in the future. Even the many mammal fossils he had already discovered were represented only by a few disarticulated bones presented in glass cases or embedded in slabs, as they had been found. Marsh's biographer Charles

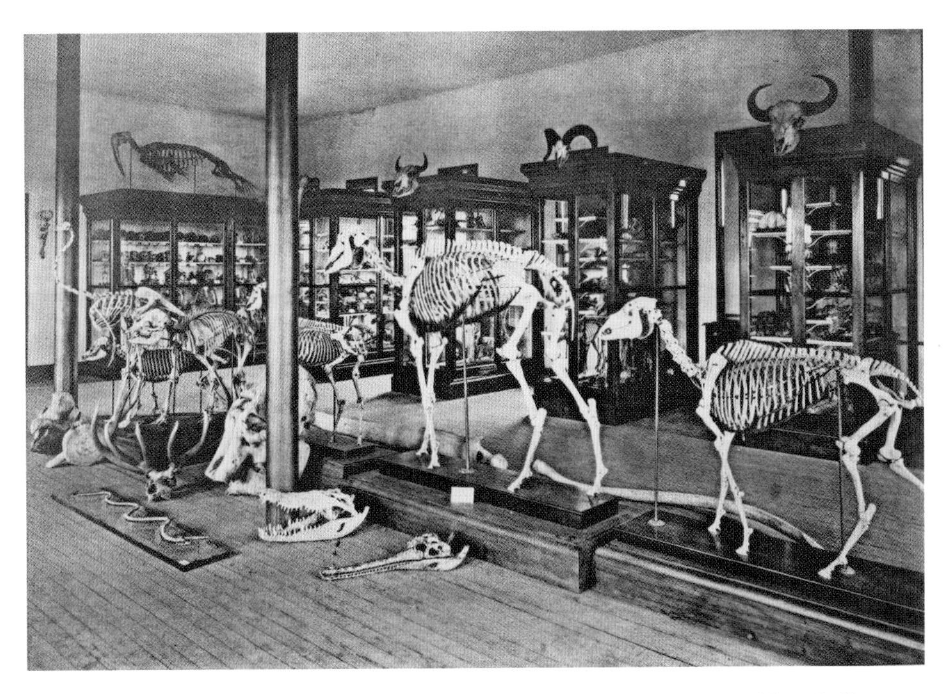

Horses and a parade of other skeletons—including rhino, elephant, and crocodilian skulls—in the first museum building.

Schuchert, who was an invertebrate paleontologist, first saw that display in 1892, when it had been somewhat enlivened by the addition of a few plaster and papier-mâché models, and he felt "bound to conclude that the show made by Marsh for the public . . . was wholly unworthy of the Marsh Collections." It conformed, however, to the museum standards of the day: "At that time, no paleontologist would have committed the sacrilege of showing the bones themselves mounted in a life-like attitude, or even modeling the missing parts in plaster. No! The bones must be kept apart forever in drawers, or laid away on padded shelves, so that the paleontologist alone might handle them and inspect every joint in the most minute detail." The choice may also have suited Marsh's possessive and somewhat secretive psychology as a collector. His motto, Schuchert later wrote, might have been, "What I have, I hold."[6]

Even so, the museum offered plenty to amaze and delight the public, including "the great bones of the Otisville mastodon," an extinct Irish elk with antlers more than thirteen feet across, a Japanese giant salamander a yard or more in length, a walrus from Alaska, and crabs displayed, as the *Scientific American* reporter put it, in "all their natural grace . . . evidently as ready as ring politicians to seize plunder with claws that will not relax."[7]

The Warm Glow of Other Peoples' Envy

From this beginning, growth of the collections rapidly accelerated. In 1877 alone, *Stegosaurus, Apatosaurus, Allosaurus,* and *Atlantosaurus* rose up from their burial places and made their way to a new life at the Peabody. In their attic refuge, the anthropological collections began to feel the squeeze. Even mineralogy must have felt diminished: "Not only the whole first story, but also cellar and attic, are filled with fossil bones," a German visitor, Karl Alfred von Zittel, reported on touring the museum in 1883, just six years after it opened. "Long rows of piled-up boxes contain the paleontological treasures. Only a very strict order makes it possible to find every thing in these crowded rooms."[8]

In Marsh's laboratory, he reported, perhaps twenty-five huge, horned skulls of Dinocerata awaited examination, and in the adjacent room the patients included "a complete skeleton of the curious *Brontosaurus,*" a *Stegosaurus,* and part of the femur of "the gigantic *Atlantosaurus.*" Assistants hurried about "preparing and combining the objects which so often arrive in fragments." Lithographers found space amid the specimens to make their plates and reconstruct these fabulous creatures in print. In a separate building, a preparator "forms casts of the finest specimens, and afterwards these casts are sent with the greatest liberality to American and foreign museums."[9]

In a spirit of gratitude and delight, von Zittel, a professor of paleontology at the University of Munich, attributed all this activity and the "fine condition" of the Peabody Museum "to the self-sacrificing activity of Professor Marsh." But he was also plainly daunted. Writing about his visit in *Allgemeine Zeitung,* then Germany's leading daily newspaper, he described Marsh's collection of fossil vertebrates as "not less complete, and not inferior in value to the collection of the British museum in London. It is infinitely more than all the material ever seen and studied by Cuvier during his whole life."[10] Von Zittel diplomatically refrained from ranking the Peabody with the paleontological collection at the museum in Berlin. But the thought no doubt occurred to his readers.

Marsh, at fifty-two, was still in his prime, with the prospect of further great discoveries ahead. He must have quietly rejoiced on reading von Zittel's words: "A visit at the Peabody museum, under Professor Marsh's guidance, arouses very mixed feelings in a European colleague." He admired "the character of greatness and magnitude which we find in many conditions of American existence." But he also despaired a little at the uninterruptedly "new and unexpected objects" being resurrected from "the virgin soil of America." It left him, he wrote, with "the disheartening conviction, that, whereas the time of great discoveries has begun in America, it is over in Europe."[11]

It was just eighteen years since Marsh's question to Dana about the future of the scientific collections at Yale: "Should not those of New Haven soon be as extensive" as those in Berlin? All Marsh lacked now was the completed museum building in which to house these assembled treasures. In 1882, the Peabody trustees solicited bids to build the central structure of J. Cleaveland Cady's original architectural design. But the projected cost of $214,772 proved too much, and it seemed to move further out of reach in the years thereafter. Marsh was gradually spending down his own inheritance on his relentless collecting, and he became a little desperate about finding a donor to step forward and house his many discoveries. Appeals for help to make the Peabody Museum whole went out. A museum keeper just down the road in Bridgeport, P. T. Barnum, pleaded poverty.[12]

Marsh surely knew that completing Cady's handsome design would have important symbolic value beyond the simple need for space. The museum would then extend along High Street for half the length of the Old Campus and anchor science, perhaps permanently, at the heart of the university. Dana and Marsh had earned worldwide reputations for their contributions to science. They had inherited Benjamin Silliman's intellectual legacy and abundantly expanded it. They had earned a commanding presence in the university. But what the Peabody really needed now was Silliman's almost magical ability to shake loose enough funding.

It never happened. Instead, the Peabody remained for decades partly built, with "a clumsy asymmetry that it was never intended to possess," according to one architectural historian.[13] By itself, that awkwardness might not have doomed the building. But in the new century, as the expanding university looked to develop residential colleges and to strengthen science, paradoxically, by moving it to a place on the periphery of the campus called Science Hill, the Peabody Museum land would inevitably become a target.

Mr. Marsh Builds His Dream House

George Peabody had left his nephews $30,000 apiece to build themselves homes, and in 1876, O. C. Marsh put J. Cleaveland Cady, the architect of the Peabody Museum building, to work designing an eighteen-room mansion on a 6.8-acre property at the top of Prospect Hill, just up from the Yale campus. Marsh filled the house with his collections—not fossils, which remained at the Peabody, but paintings, pottery, stuffed animals, orchids from the greenhouses out back, cloisonné metalwork, Japanese scroll paintings, and an abundance of oriental rugs. The construction and furnishing were finally completed in time for Marsh to take up residence in 1883.

Marsh's house became best known for the octagonal reception room, dubbed the Wigwam, where he often entertained students and visiting dignitaries. One guest described it as "a sort of trophy room filled with mementos and treasures from all over the world. Here a scalp or a pair of buckskin leggings, or a frontiersman's pistol would recall some incident of the west and Yale seniors became small boys again, listening to tales of Indian savagery, or of hairbreadth escapes from stampeding buffalo."[a]

Stripped of much of its ornate past, the house now serves as an office and classroom building and is listed on the National Register of Historic Places.

Marsh stocked his ornate house in New Haven with memorabilia from his expeditions and spun stories about them for students and celebrated visitors at dinners there.

CHAPTER 12

In the Shadow of O. C. Marsh

During vacation a young elephant was received from N.Y. in a very odorous state, but I have prepared a very fine skeleton from it. . . . A large ostrich arrived a few days ago & is now soaking.

A. E. VERRILL to O. C. Marsh, 1871

Late in October 1873, three fishermen out for herring on Conception Bay in Newfoundland spotted a bulky object floating on the surface of the water. Curious, they rowed their dory closer, and one of the men prodded it with his boat hook. "Instantly, the seemingly dead mass became animated," Rev. Moses Harvey, a clergyman in nearby St. Johns, later wrote. "It reared itself above the waves, presenting a most ferocious aspect, and displaying to the horrified fishermen a pair of great eyes, gleaming with rage, and a horny beak, with which it struck the gunwale of the boat. The next instant a long, thin, corpse-like arm shot out from the head, with the speed of an arrow and coiled itself around the boat. It was immediately followed by a second arm, much stouter but shorter, and both, in some mysterious way, glued themselves to the boat, which presently began to sink. The terrible monster then disappeared beneath the surface, dragging men and boat with it."[1]

Jules Verne had described just such an attack by "an immense cuttle-fish" with "enormous staring green eyes" in *Twenty Thousand Leagues under the Sea,* recently translated into English. Victor Hugo had also lately dwelt on the "sea vampire," or devil fish: "A glutinous mass, endowed with a malignant will; what can be more horrible?" Harvey's dramatic account, headlined "How I Discovered the Great Devil-Fish," may thus have been somewhat derivative, not to say fictionalized. "The water was pouring into the boat as it sank lower and lower," he continued, "and in a few seconds all would have been over with the unfortunate men." At the

The Reverend Moses Harvey collected the giant squid tentacle at right and sent it to Verrill. The beak at left came from another giant squid specimen.

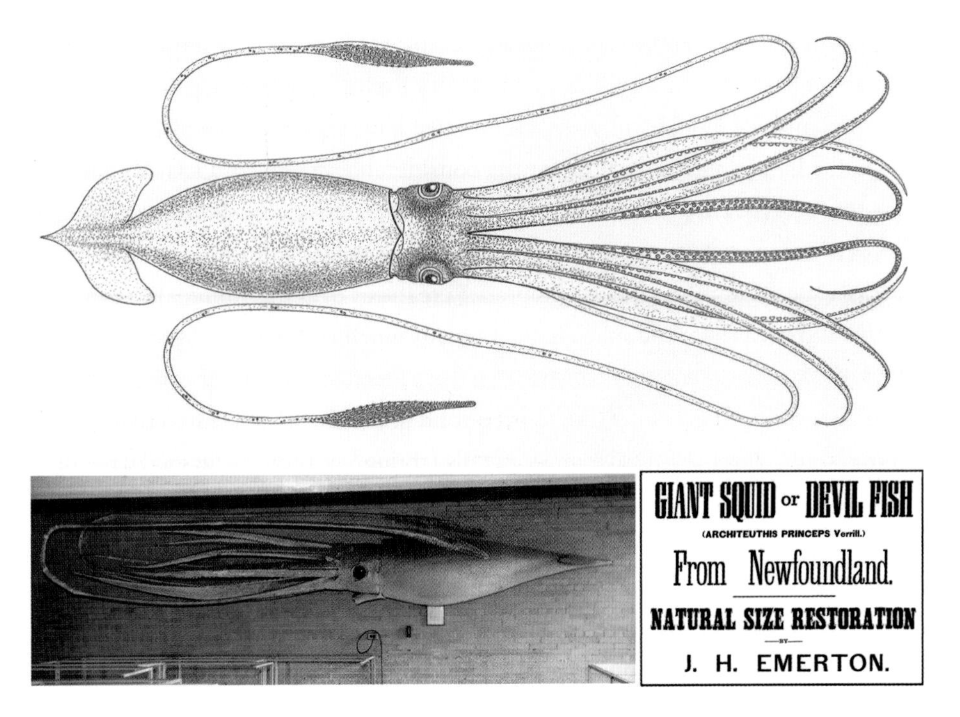

Verrill's illustration of the giant squid along with a model that once hung in the museum.

last moment, the helmsman grabbed "a small tomahawk . . . dashed forward, and with two or three quick blows cut off both arms as they lay over the edge of the boat. The creature did not attempt to renew the fight." Back on land, one of the fishermen tossed the shorter arm outside his door, where dogs ate it. But "the long thin arm he carefully preserved." It was nineteen feet long, and he had the odd idea that it might make a mooring line for his boat.[2]

Harvey bought it instead and "decided that the interests of science would be best served by sending my specimen" to Yale professor Addison E. Verrill, "an eminent naturalist." In Verrill's laboratory, it became part of the evidence for the first detailed scientific description of the giant squid, named *Architeuthis dux* a few years earlier by the Danish zoologist Japetus Steenstrup. Verrill "made an exhaustive study of it, and described and figured it in numerous scientific periodicals; so that its fame spread over the whole world," Harvey continued.[3] The important result was that Verrill turned one of the world's most mythologized monsters into a real animal and, as Harvey put it, "the fabulous 'Kraken' of the old naturalists has been thrown into the shade." In the spirit of continued collaboration, Harvey sent Verrill a sample of a sixty-two-foot-long tapeworm freshly extracted from a salmon.

relationship gone bad, and in 1864 when Dana was looking to hire Yale's first professor of zoology, Verrill made his escape to New Haven. He had been a devotee of Dana's *Manual of Mineralogy* from the age of twelve, and the two men also had common ground in their coral research.[7]

At the Peabody Museum, Verrill soon began to build a proper collection across the breadth of the animal kingdom. He had a budget then of just $150 to $300 a year for acquisitions, but he also worked to persuade private collectors, other museums, and menagerie keepers to donate interesting specimens and inconvenient carcasses. "During vacation a young elephant was received from N.Y. in a very odorous state," Verrill noted at one point, "but I have prepared a very fine skeleton from it. I believe it is all complete. A large ostrich arrived a few days ago & is now soaking."[8] Before the Peabody had its own building, Verrill's laboratory was on the ground floor of the former Trumbull Gallery, then renamed the Treasury Building, at the center of the campus. The receipt of specimens in various stages of ripeness was warmly noted by college administrators just upstairs.

Coastal Communities

Verrill was also collecting on his own and with his students, seining and trapping in the waters of southern New England, mainly around New Haven. That effort vastly expanded beginning in 1871, after Spencer Fullerton Baird, assistant secretary of the Smithsonian Institution, persuaded Congress to investigate an alarming decrease in coastal and interior fisheries. Baird himself took charge of the new U.S. Fish Commission and he set out to study not just commercial fish species, an assistant later wrote, but also "the histories of the animals and plants upon which they feed . . . the histories of their enemies and friends and the friends and foes of their enemies and friends, as well as the currents, temperatures, and other physical phenomena of the waters in relation to migration, reproduction, and growth."[9] That summer, and continuing every summer until Baird's death in 1887, Verrill served as chief scientist for the Fish Commission. Over the winters he studied and described the invertebrate specimens commission scientists had found. The Smithsonian had first claim on these specimens, but the duplicates would reside at the Peabody Museum.

The Fish Commission set up its first summer station at Woods Hole, Massachusetts, in 1871, the beginning of that community's status as one of the world's leading centers of marine research. Cape Codders must have struggled at first to comprehend the newcomers in their midst. A growth on a piling that they might

At the same time Verrill was building the Peabody's vertebrate collection from scratch, Daniel Cady Eaton created the museum's herbarium and published his masterwork *Ferns of North America.*

dismiss as seaweed—or, more likely, never notice at all—was, for Verrill, "one of the most beautiful, as well as one of the most abundant, of the Hydroids," with "large clusters of branching stems, often six inches or more in height, each of which is surmounted by a beautiful, flower-like, drooping head of a pink or bright red color." But if Verrill explained that much to a curious passerby, he might then feel obliged to add that these "seaweeds" were actually animals. And that might encourage him to add that they often lose those colorful heads, not "a very serious accident, though certainly a very inconvenient one, for the mouth, stomach, tentacles, and most other organs" tended to go drifting off in the process. It dawned on Verrill one day that he might be better off just explaining to the locals that he did what he did because someone was paying him to do it. It was a lie, but the ploy made sense in Woods Hole.[10]

One result of the Fish Commission work was Verrill's first book, *Report upon the Invertebrate Animals of Vineyard Sound and Adjacent Waters, with an Account of the Physical Features of the Region.* To the lay reader, it is a largely colorless work, encyclopedic but devoid of evocative descriptions or personal anecdote. To students of marine invertebrates in the western Atlantic, though, it is a sort of bible to which they still turn when they want to know what a particular species looks like or precisely where to find it. Marine biologists Joel W. Hedgpeth and Harry S. Ladd, writing in 1957, called it "the classic work in American marine ecology,"

A Visible Woman

In 1879, when funds became available for an assistant to help with the rising tide of invertebrate specimens from the U.S. Fish Commission, Addison E. Verrill made a choice that was remarkable for that era: he hired a woman. At age twenty-three, Katharine J. Bush had no special training in science. She was the child of an accountant and a homemaker, and her education had ended at New Haven's Hillhouse High School (though her male cousins had continued on to Yale). But her job arranging, cataloguing, and labeling specimens at the Peabody opened a door into the scientific world for her, and her talent for the nuances of invertebrate taxonomy quickly became evident.

Other women already worked in science. But they were typically unpaid "volunteers" whose husbands or male colleagues took credit for their work. Verrill not only paid Bush, he also seems to have encouraged her to become visible as a scientist in her own right. In 1883, just four years after starting as his assistant, she published an article under her own name in the *Proceedings of the U.S. National Museum* about starfish and sea urchins from a dredging expedition in Labrador. She would continue to publish for the rest of her career, sometimes on her own, other times sharing a byline with Ver-

Scientific women: the Peabody Museum's Katharine J. Bush (*top*) posed in 1886 at Woods Hole with Mary Jane Rathbun (*left*) of the Smithsonian, Eloise Edwards, who was a friend, and Charlotte Bush (*far right*) of Yale's Sheffield Scientific School.

rill—testimony, perhaps, to Verrill's determination not to repeat his own unfortunate experience under Louis Agassiz.

When the Fish Commission work ended in 1887, Bush became a (modestly) paid member of the staff at the Peabody Museum. She and her sister Charlotte Bush had also begun taking courses as "special students" at Yale's Sheffield Scientific School. Charlotte would go on to a conventional life, first becoming librarian at "the Sheff" and then marrying Wesley R. Coe, a Verrill student who would later succeed him as the Peabody's curator of zoology. But in 1901, Katharine Bush became the first woman to earn a doctorate in zoology at Yale. Her thesis described three new genera and sixteen new species of tubiculous worms, collected mainly by Coe on the 1898 Harriman Expedition to Alaska. A third sister in the family, Lucy P. Bush, also worked at the Peabody, but always behind the scenes, performing editorial and clerical duties for O. C. Marsh and inevitably also crossing over into scientific work with the fossils he collected.

A doctorate did not, of course, put Katharine Bush on course to become a curator or assistant professor, nor could she do much in the way of fieldwork. Though less restrictive than foot binding, the corsets, long skirts, and parasols of the Victorian era (all evident in the photograph opposite) effectively prevented her from joining expeditions. But even so restricted, she could at least work with specimens and publish her findings.

Another female scientist also got her start in Verrill's lab. Mary Jane Rathbun, remembered now both for her size (four feet six) and her quick, dry humor as well as for her science, came to Woods Hole with her brother Richard Rathbun and both started out as volunteers with the U.S. Fish Commission. Her education had ended at high school, following the familiar pattern, while he'd continued to Cornell and Harvard. But sorting specimens in Verrill's lab under the tutelage of Sidney I. Smith, an expert on crustaceans at the Peabody, Mary Jane Rathbun soon also became an expert on crabs, shrimps, and other crustaceans. Her brother went on salary with the U.S. Fish Commission, working in the field and at the Peabody until 1881. Then he moved to Washington, D.C., where he would eventually become director of the U.S. National Museum and assistant secretary of the Smithsonian Institution. Mary Jane Rathbun followed him, working as a volunteer for many years before eventually being allowed a modest clerkship.

Verrill characteristically left no hint of how he came to flout the sometimes fierce prejudice against women in science. Maybe there was a tradition of strong women in his family (a great-granddaughter would become a bullfighter under the name la Diosa Rubia, the Blonde Goddess).[a] Or maybe he simply recognized that hiring an underpaid female scientist could be an efficient way to eke out his research budget. In any case, Bush's "faithful assistance" over a period of more than thirty years was one reason Verrill accomplished twice the work of many other scientists in describing new species and publishing scientific papers. On his death, Wesley Coe wrote that Bush's "accuracy and ability are reflected in nearly all of Professor Verrill's publications during that period."[b]

Bush never married, and according to painstaking research in the 1970s by Jeanne Remington at the Peabody, she made her solitary home in "a series of rented rooms" around New Haven. No hint of her personality survives, except from an episode in 1909 when she published a paper suggesting that a Smithsonian researcher named Paul Bartsch had stolen her unpublished work on a series of beautiful turret- and turban-shaped shellfish. That was, of course, what male scientists did, and the episode seems to have tainted her reputation more than his. ("My guess is that Katharine was rather emotional and easily upset," a Harvard researcher wrote in a letter to Remington almost seventy years later. It was the standard put-down of women in the workplace, though written by a woman.)[c]

In 1912, Bush received one of the only public acknowledgments of her work, a brief mention in a book, *The Part Taken by Women in American History,* as "one of the noted scientific women of America."[d] Soon after, some unrecorded illness disabled her, requiring the Peabody to hire two assistants as her replacements. She soon took up residence in the Hartford Retreat, a sanitarium, where she remained until her death in 1937. There was no obituary.

mainly for its treatment of what Verrill termed "distinct assemblages of animal life, which are dependent upon and limited by definite physical conditions of the waters which they inhabit."[11]

It may seem obvious now that different physical conditions—for instance, rocky, sandy, or muddy shorelines—would foster different communities of marine life. But the idea of ecological communities, or ecosystems, was then unknown. Verrill came tantalizingly close to establishing this fundamental concept in biology. He didn't just describe habitats and how the species within them interact; he also described microhabitats within those larger habitats—for instance, how creatures live around the bases of plants or under rocks. But he never made the leap to generalizing from his facts, never put forward any grand theories. Thus Karl

August Möbius, who was doing similar work on marine organisms in Germany's Bay of Kiel, earned the credit for the idea of biological communities in a book published just three years after Verrill's work, which he cited merely as supporting evidence.

Verrill stayed focused on his collecting. In Gloucester, Massachusetts, the Fish Commission somehow managed to persuade the local fishermen to compete to find prize specimens, presenting Verrill with what would eventually turn out to be about fifty new invertebrate species. Verrill also improved his catch by tinkering with the machinery. He introduced the use of a wire rope for dredging and devised innovations like "trawl-wings" for catching free-swimming creatures along the bottom. A "mop-tangle" Verrill invented for collecting starfish was still being used decades later by New England shell-fishing boats to remove these voracious predators from oyster beds.[12]

Under Baird's direction, Verrill gradually moved the collecting farther out to sea. He soon recognized that the Gulf Stream fueled local fisheries by carrying large quantities of organisms into the waters off New England. So the Fish Commission designed its own thousand-ton vessel, the *Albatross,* to collect on transects there. It would eventually work as far south as the Carolinas and to a maximum depth of twenty-eight hundred fathoms, or more than three miles down. Fish Commission biologists later boasted that their quiet and methodical work had resulted in the discovery of forty-seven new genera and 147 new species of deepwater fishes, outnumbering equivalent discoveries by the HMS *Challenger,* which was becoming famous in the same period for its four-year round-the-world oceanographic expedition.

Specimens in Lieu of Salary

Verrill continued to work on the resulting invertebrates at his microscope until 1908, long after his separation from the Fish Commission. (He had been distracted for a time by two collecting expeditions in Bermuda, a stint updating all the zoological definitions in *Webster's International Dictionary,* a tour as Connecticut's state parasitologist, and innumerable other assignments.) The first pick of specimens then went off, belatedly, to the Smithsonian. The duplicate group was almost as good because Verrill typically identified syntypes, or groups of individuals, as the defining examples of an invertebrate species, with the syntypes for any common species represented equally in each collection. But now Verrill proposed to sell the duplicate collection, preferably to the Peabody Museum, where it had become the soul and substance of the invertebrate zoology division, but otherwise to the highest bidder. He printed a pamphlet advertising the collection as including more than fifteen thousand lots representing three thousand species, comprising nearly all known invertebrates from the Atlantic coastal region, and some still undescribed. Many came from shallow waters, but most were "from great depths, far out at sea" and unlikely ever to "be obtained again except by similarly costly government expeditions."[13]

Profiting from government-financed collections might seem now to be ethically questionable. At the time, though, the *New York Times* treated the cost of gathering the specimens, which it estimated in the hundreds of thousands of dollars, as no more than a bragging point. But for Yale, Verrill's proposed sale represented a grave breach of unspoken gentlemanly and scientific codes. The old notion, inherited from the British, was that naturalists should be people of independent means, on the order of Charles Darwin or the Duchess of Portland. Yale and most other institutions took advantage of this assumption, and people like Verrill, who had no such means, had to scramble for a living. He earned $1,200 in his first year at Yale, which was double what Agassiz had paid him. But even as Verrill's salary slowly increased over the years, it "was never sufficient to support his family," according to George E. Verrill, the eldest of his six children, and it topped out after more than thirty years at just $3,750.[14] Verrill griped about it from the 1860s onward, resenting the need to take time away from his research to undertake more lucrative work as a consultant or writer.

Even worse, the U.S. Fish Commission, which consumed two or three months of his working year in field research alone, paid him no salary whatever (his white lie to curious bystanders in Woods Hole to the contrary). His apparent understanding with Baird was that the duplicate specimens would stand in lieu of salary. So selling the collection seemed to Verrill to be simply due recompense. In any case, his pamphlet elicited a bid from the American Museum of Natural History for $22,500. The Peabody trustees gulped and sent a plea to the Yale administration noting that the collection was "of great value and hardly to be replaced" and, not incidentally, that losing it would represent "a most serious blow to the prestige of the University." Verrill was thus able to extract an $18,000 deal to leave the Fish Commission duplicates and some of his private collections at the Peabody. A note followed soon after from Edward S. Dana forcing him into retirement, though in the "hope you will not misunderstand the spirit which has prompted this action."[15]

It was an unfortunate end to his time at the Peabody, especially given the acknowledgment in the same letter of "how brilliant your career in science has been, and how large a debt the University and Museum owe to you." Verrill somehow negotiated office space in another building on campus and of course continued to work on his invertebrates. But this ultimately led to an even more painful break with Yale. In 1925, when he was eighty-six years old, Verrill traveled to Hawaii with his son's family, and in his absence, someone unwisely gave a student permission to use his office. Verrill somehow got wind of it, and all the perceived slights and buried anger from his sixty-year association with Yale came boiling to the surface: "I can hardly believe that such a thing can be true," he wrote to the head of the zoology department. "If so it is a very unjust and cruel thing to do." He detailed all the specimens, drawings, and articles that might be jeopardized by a student's carelessness and concluded: "If I am to be turned out of doors by Yale Univ. I cannot think of sending anything there," meaning the remainder of his private collection of specimens. "Harvard or the N. York. Amer. Museum; or the National Museum would treat me with far more consideration I am sure."[16]

Grace and Beauty

Verrill died a year later and was soon largely forgotten by the outside world. George E. Verrill later wondered (to the horror of invertebrate zoologists everywhere) whether his father might have become "more widely known" had he "devoted his life to mining and geological work, including the examination of prospective oil fields." But the truth is that Verrill was enraptured with the living world. He never quite articulated this driving motivation—doing so might have seemed like one more distraction from the facts, one more unnecessary generalization. So the words of a colleague and friend who was also an invertebrate zoologist must make the point by proxy: "There is a singular delight," the Smithsonian's William H. Dall once wrote, "in taking these delicate and almost microscopic animals and putting them under a strong glass, seeing the tiny heart beat, and blood circulate and gills expand, counting the muscles and blood vessels, and almost the tiny disks that form the blood and to know that you are the first that has penetrated these mysteries and are perhaps the only one who ever will, and that all your notes and drawings and observations are so much solid knowledge added to the worlds store, so much more testimony to the power and grace and beauty of the Infinite."[17]

Verrill, his son conceded, had done "the work he liked best, and was happy in doing it, which after all is the main thing, and he gained a world wide reputation in his own chosen field of science."[18] Verrill spent the last summer of his life in the place he had come to love most, a six-acre island made of pink granite two and a half miles off Branford, Connecticut, which he had colonized and made habitable for his family. Outer Island was close enough to Yale and his collections, but also far enough away. Verrill of course had an office in the house he had built, and no doubt spent too much time at his desk there working (and now and then dozing). But it was also a place where he could step away for a moment to gaze down at the teeming invertebrate life in the tidal pool below his veranda, or look out across Long Island Sound to the open sea. And then, before too long, a promising specimen would call him back to his true life—to the wonders beneath the lens of his microscope, waiting to be discovered and meticulously described.

CHAPTER 13

The Prince of Bone Hunters

I would not be afraid to tackle one now that weighed ten thousand.

JOHN BELL HATCHER to O. C. Marsh, 1891

Late in 1888, John Bell Hatcher, bone hunter, was traveling alone through southeastern Wyoming. He looked like—and was rumored to be—an excellent poker player, his high forehead hidden beneath an oversized white Stetson, his eyes deep set, the brows slanted down to the outsides, his lips thin and expressionless. His face was "inscrutable," said an acquaintance, "and you never knew whether he had a bob-tail flush, or a full house."[1] Hatcher's adventures as a bone hunter—his daring, his determination, his unshakeable integrity, and even his ability to handle playing cards or a gun—would make the pulp fiction Yale heroes of a later day, the Dink Stovers and Frank Merriwells, look like pale shadows of men.

Poker was, among many other things, a handy means of striking up friendships and trading small talk as he traveled through the Deadwoods and Laramies of the developing West. In Douglas, Wyoming, one day, the talk turned to a huge skull some cowhands had found three years earlier. A rancher named Charles Guernsey described it as "sticking out midway of a bank on one side of a deep dry gulch" with "horns as long as a hoe handle and eye holes as big as your hat."[2]

Guernsey and his cowhands had tried to dig it out of the sandstone bluff, with the result that it tumbled to the bottom of the ravine, almost taking Guernsey down with it. He had carried away only a fossil horn core—the horn minus its eroded exterior sheath. At Hatcher's request, Guernsey sent it to New Haven for inspection that winter. Its huge size—almost twenty inches long and eight inches across at the base—made Hatcher and Marsh both question Marsh's recent attribution of another fossil horn core to a species of ancient bison. Marsh was "im-

John Bell Hatcher in about 1885. Equipped with "powers of vision . . . at once telescopic and microscopic," he discovered some of Marsh's most spectacular fossils.

mediately possessed with a burning desire" to obtain the skull and identify the geologic strata from which it came.[3] Early that May, under heavy pressure from Marsh, Hatcher left his wife, Annie, who had lost their first baby just two weeks earlier, and headed to the scene, some thirty miles north of Lusk, Wyoming.

"The big skull is ours," he wrote just three days later, with characteristic efficiency.[4] The fragments filled four large boxes and weighed about eleven hundred pounds. It was a small sample of the much larger discoveries lying just ahead as Hatcher hunted down some of the most remarkable creatures in paleontological history.

Enormous Undertakings

Hatcher had grown up in Iowa and attended Grinnell College for a year before transferring to Yale. He earned money for college working as a coal miner, and the fossils he saw as he worked stirred his interest in paleontology and eventually brought him to Marsh's attention. Within days of Hatcher's graduation in 1884, Marsh sent him, with funding through the U.S. Geological Survey, to start an apprenticeship under Charles H. Sternberg, founder of a Kansas empire in commercial fossil collecting. But Hatcher had strong ideas and an astounding sense of self-reliance. He thought Sternberg was overlooking valuable specimens and could save others "by taking more pains" in digging them out and packing them.[5] One day, after Sternberg had repeatedly failed to pacify the troublesome owner of the land where they were digging, Hatcher discerned the true nature of the difficulty and hired the landowner as an assistant. Then the two of them moved off to the opposite side of the ravine from Sternberg, and Hatcher's prodigious career as an independent collector had begun.

The first report Hatcher sent to Marsh, just four days later, showed the site carefully divided into sections and mapped out, much as a modern paleontologist would do, and with the position of each of the sixty-seven bones he had removed from one section precisely located on a separate diagram.[6] Over the next few years, he repeatedly proved his worth by collecting in all seasons and circumstances, in the process developing an excruciating and chronic case of inflammatory rheumatism but also sending hundreds of specimens of *Brontotherium* back to New Haven. These were huge mammals from the late Eocene, rhinolike in appearance but as big as African forest elephants, and with a pair of horns like a goalpost above the nose. Marsh named them *Brontotherium,* or "thunder beast," from an Indian belief that the hoofbeats of these enormous animals were the source of thunder. He wanted as many specimens as possible for a monograph he hoped to publish, like the one he had recently completed on the Dinocerata.

In 1887, Hatcher found time to court and marry. But given his unstoppable drive as a paleontologist and Marsh's eagerness to take full advantage of it by keeping him in the field year-round, Anna Peterson Hatcher, a Swedish immigrant, must have come to the marriage with a singularly forgiving nature. Fossils were Hatcher's true love. They were the remains of creatures that had been dead and extinct for millions of years, and yet no woman ever faced a more formidable rival. Annie kept house, often on her own, in a Nebraska town aptly named Long Pine.

Still in his midtwenties, just four years out of college, Hatcher now demonstrated the combination of skills that would win him recognition as "the best and most successful palaeontological collector" America ever produced. First among these skills was a genius for reading rock. Hatcher could determine the direction in which a current had flowed roughly 66 million years ago by noting the sand built up on one side of an exposed bone and the fossilized plant stems and leaves that had sunk and been buried after being caught up in the resulting eddy on the opposite — downstream — side.[7] That sort of clue often led him to bones in places where a less observant paleontologist might never have bothered to look.

His "marvelous powers of vision" were "at once telescopic and microscopic," the paleontologist W. B. Scott wrote, and he displayed both with "a dauntless energy and fertility of resource that laughed all obstacles to scorn." Wyoming's Lance Formation was strewn with the remains of diminutive Cretaceous mammals. But it could take a man half a day of conventional searching, bent at the waist or on hands and knees, to come up with a single tooth. Hatcher turned to screening as a more efficient method, and at one point he had thirty wagonloads of quarry sand waiting to yield up their mammal parts. Then it occurred to him to inspect the contents of nearby harvester ant mounds, running them through a small flour sifter. Ants of this species spread out across the landscape collect-

This box, deliberately kept as it was on receipt, contains the hipbone of a duckbilled dinosaur. It was among the last specimens Hatcher shipped in 1892 before leaving Yale to work with a Marsh rival.

ing seeds to eat and also pebbles and other small objects to insulate the mound. Hatcher found he could recover as many as three hundred teeth and jawbones from a single mound. Thereafter, in areas that lacked harvester ants, he would introduce a nest of them and check on their work a year or two later. One such introduced nest yielded thirty-three fossil mammal teeth.[8]

Hatcher was also unequaled in the care and the mechanical aptitude he brought to the delicate business of extracting the much larger specimens he found. "Out here, mechanical facilities are few and such large undertakings are only accomplished by the hardest work and perseverance under many difficulties and disadvantages," wrote Charles E. Beecher, a Peabody paleontologist who was working with Hatcher in the summer of 1889. And yet Hatcher managed to extract "huge and weighty objects from difficult positions," another writer remarked, in the face of "great physical risks and even death, when alone, far from human companionship." In one case, "by a skillful arrangement of levers," he extracted "a block of rock weighing nearly a ton without the assistance of other men."[9]

By the end of July 1889, Hatcher had two more gigantic skulls, each weighing about three thousand pounds, counting the stone in which they were embedded. "They are certainly most marvelous creatures," Beecher wrote, "with their scalloped and cornuted frill, their big horns and extravagant beaks. These specimens will make a wonderful and startling exhibit."[10] But first Hatcher had to haul them intact out of steep ravines, crate them for long-distance travel, and deliver them by wagon thirty or forty miles across rough country to the railroad at Lusk, a three-day trip.

"But I have an abundance of 'self conceit' & it does not frighten me at all,"

Hatcher wrote to Marsh, "& you may be sure that we will get them out & packed all o.k." ("It would make your hair stand on end, to see the terrible places where Hatcher will take a heavily loaded horse," a collector who worked with him remarked. "Places you would not imagine it possible for a horse to go at all.") Two years later, working in continuous rain and snow, he extricated a 6,850-pound skull and shipped it in a crate that was ten feet long and five wide. "I would not be afraid to tackle one now that weighed ten thousand," he remarked afterward.[11]

The skulls he was finding, measuring six or eight feet in length, belonged to *Triceratops,* as Marsh named the genus in a November 1889 article. That article gave a brief nod to Hatcher in its final sentence, an unusual gesture for Marsh, who otherwise recognized Hatcher's talent mainly by sending him to work in northern regions for nine months of the year and southern ones for the other three. As with all his collectors, he also routinely kept him on a short budget (though Hatcher is said to have supplemented this with poker winnings). Hatcher's letters from that period are full of reports of amazing discoveries alternating with pleas for past-due payments. Though he was ostensibly earning $300 a month for wages and incidental expenses, he calculated that he cleared just $80 in July 1889. August was worse. After paying to freight two huge skulls back to New Haven, "I will be about $8.00 worse off than if I had not dug bones at all this month." In case Marsh did not get the point, he added, "In other words, I pay $8.00 for the privilege of working a month and furnishing an outfit [the wagon and horse team] that has cost me over $500." He had not seen his young and grieving wife since May, and he wrote, "I have been disappointing people all summer."[12]

Hatcher wanted a chance to become not just a collector but a scientist describing his own finds. He also yearned for the privilege of leading a normal domestic life, at least for half the year. ("A letter from Long Pine tells me we have another young bone-hunter at our house," Hatcher wrote Marsh when Annie Hatcher gave birth in August 1890 to their second child. "I hope he will live longer than the other one did.")[13] Hatcher asked repeatedly for a permanent salaried position, preferably at the Peabody or at the Smithsonian's Natural History Museum. Either would have allowed him to work with specimens he had collected for the U.S. Geological Survey. Failing that, he asked Marsh for a recommendation to help him obtain such a position elsewhere. In 1891, Marsh agreed to a contract that gave Hatcher a $2,000 annual salary as assistant in geology and stipulated no more than six months a year in the field. But this was a fiction. The two finally ended their connection in 1893, after funding from the U.S. Geological Survey ran out. Hatcher moved to a position as curator of vertebrate paleontology at Princeton, where he worked under a Cope ally, W. B. Scott. Later, he switched to the

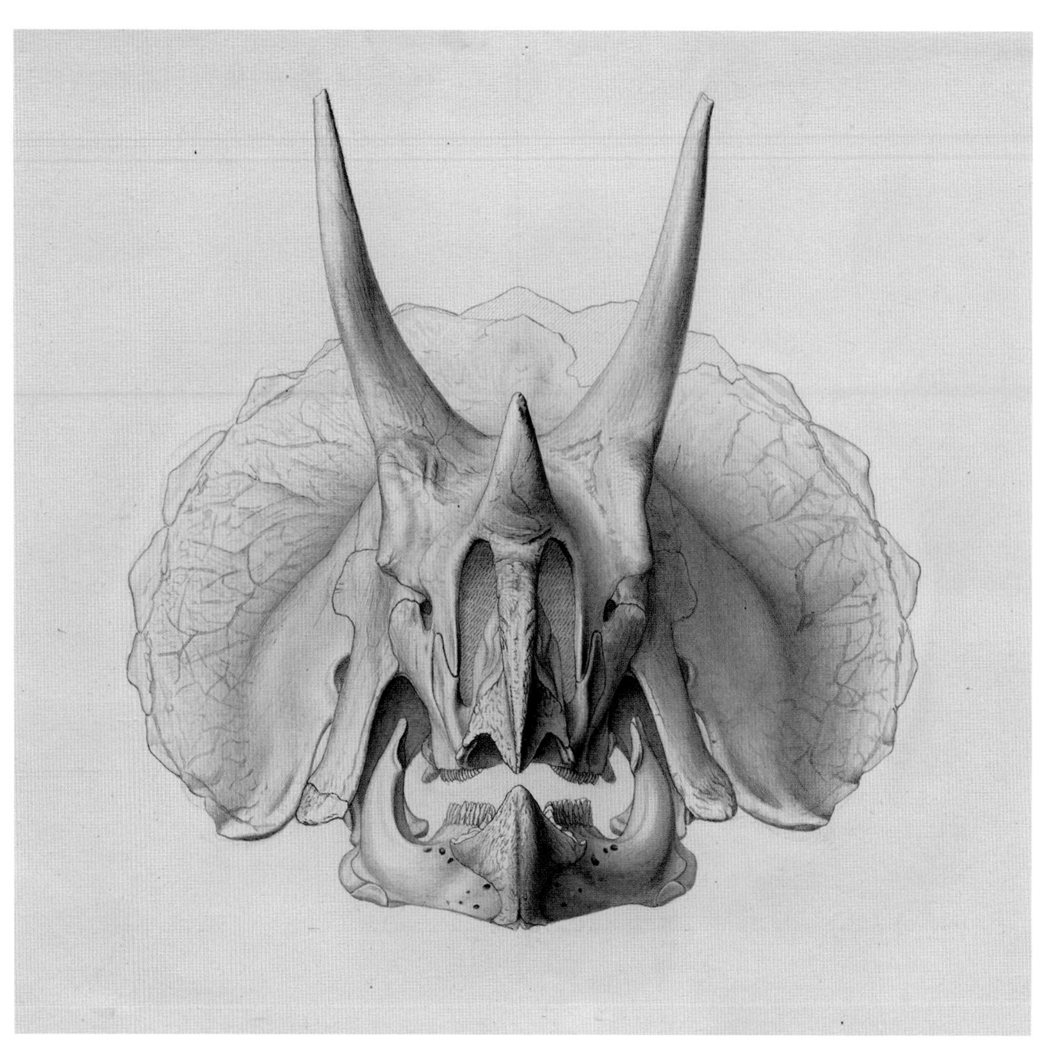

A museum artist's original drawing of the skull of *Triceratops prorsus,* discovered by Hatcher in 1889 and named by Marsh.

Carnegie Museum in Pittsburgh. But his Yale roots ran deep, and his connection to the Peabody would be renewed long after his death, largely through his work at the other end of the hemisphere in Patagonia.

Hatcher made three difficult expeditions there in the 1890s in search of Mesozoic mammals. He traveled on horseback, alone or with one or two companions, for hundreds of miles through severe winter weather, ignoring repeated warnings from locals that he should not go. But "we had tented it for many years on the wind-swept plains of Wyoming, Montana, and the Dakotas," Hatcher wrote, "often with the thermometer far below zero, and had no uneasiness as to our ability to survive successfully whatever blizzards Patagonia might have in store for us." He underestimated Patagonia. At one point on the second expedition, he crawled into his tent crippled by a combination of fever and rheumatic swelling of arms and legs but expecting nonetheless to resume his journey in a day or two. "Such was not to be the case, for the rheumatism, together with the accompanying fever, rapidly increased in severity and soon spread to my hands, feet, neck and left hip. For six weeks I was absolutely helpless and unable to shift myself in bed or attend to my most trivial wants." Another member of the expedition kept him alive. Then they went back to collecting.[14]

On another expedition, Hatcher set out alone with his horse on a 125-mile trip through horrible weather to ensure that the tons of fossils he had collected were properly loaded into a ship bound for New York. He was an experienced horseman, but during a break on the third day, his horse caught a hoof through the reins. As Hatcher stooped down to free it, the startled horse jerked its head violently upward then down again, driving "the broken shank of the Logan bit with which the bridle was fitted . . . through and under [my] scalp," tearing the skin loose "over a considerable area." Hatcher used his hand in a sort of comb-over to lay his scalp flat again on his skull, bathed his head in cold water, then mounted his horse and rode on with blood pouring down to his chest and saturating his clothes. Feeling faint after a few hours, he dismounted, wrapped a couple of handkerchiefs around his head, rammed his Stetson down around them, and lay down for the night with his saddle for a pillow. When he reached a ranch the next day, he asked for food and some soap and water to clean his wound, but the cook ("a surly Italian") told him to move on. Hatcher showed the money he was ready to pay for help, to no avail. But he had a way of getting what he wanted. (Once, in Punta Arenas, he and another paleontologist saw a horse dealer beating a horse "with the loaded end of his quirt." Hatcher drew his revolver and said, "Beat that horse once more, and you'll be a dead man." He did not need to shoot.) Now he simply pushed past the cook ("this 'dago'") and sat down in the kitchen to a meal of bread and cold mutton. He built up the fire in the stove, made coffee, and bathed his head in warm water long enough to loosen the Stetson and handkerchiefs. When he was done, he left the absent proprietor a note offering to pay for everything on his return trip. Then he rode twenty-five miles onward and spent another cold night on the ground. When he eventually arrived at his destination, Hatcher secured his specimen crates, which were of course not being handled to his satisfaction. He also took his suppurating head wound to the local doctor, who recommended bleeding. Hatcher chose to treat himself instead, successfully enough.[15]

All this naturally made an impression on the Patagonians, and an American oil executive who traveled through the region in Hatcher's aftermath recounted stories of his exploits, perhaps embellishing his reputation. Writing decades later in the *Atlantic,* James Terry Duce described Hatcher as "a rather subdued-looking American scientist" who taught lessons in poker in "every hamlet from Bahia Blanca to the Straits," and "as a rule the loose change of the community passed on to the bone hunter to be spent on science." The gambling ended when "the whole countryside dropped in to exact revenge" shortly before Hatcher's return home. "The stacks of pesos in front of Hatcher climbed up and up until he was almost hidden behind them; the whistle of the steamer sounded down the harbor. Hatcher announced that he must go. Someone suggested that they would not let him. He picked up his gun and his pesos and backed through the door with a 'Good night, gentlemen!' No one made a move. The wind whooped round the eaves and Patagonia went back to its sheepshearing with a wry smile on its face."[16]

Hatcher's own memoir of his time in Patagonia, *Narrative of the Expeditions: Geography of Southern Patagonia,* makes no mention of poker, though it is otherwise a thrilling account. He dedicated it, on publication in 1903, not to his poor, patient wife, Annie, but to O. C. Marsh, "Student and Lover of Nature," the bitter feelings from his years in Marsh's employ having faded.[17] In the decade after leaving Yale, he published forty-eight scholarly papers on fossils and geology, and his Patagonia discoveries became the heart of Princeton's vertebrate paleontology collection. At the Carnegie, he created the first museum mount of a complete sauropod, *Diplodocus carnegiei,* with a copy mounted soon after at the British Museum.

In 1985 Princeton decided to close its paleontology program and, after considering applications from the Smithsonian, the American Museum of Natural History, and other major institutions, elected to donate its entire collection to the Peabody Museum. When news reached him in the afterlife that his own specimens would end up in O. C. Marsh's museum, Princeton's W. B. Scott no doubt died again. But Hatcher might well have been content to see so many of his brilliant discoveries united in one place, where they continue to be studied by scientists from around the world. Hatcher himself died of typhoid fever at the age of forty-two, leaving his wife too soon, as he had always done, and with four young children. He was buried in Pittsburgh in a grave that was, until recently, unmarked.

Marsh (left) and his archrival Edward Drinker Cope sometimes seemed fixated as much on each other as on their fossils.

lish this volume, said Osborn, "was one of the great disappointments of Cope's life."[6]

Cope's life was full of disappointments at that point, as his mining stocks and job applications all went amiss. In 1886 he was forced to rent out the family home on Pine Street and move with his wife and daughter to a small apartment around the corner. Of his collectors in the West, he wrote to Osborn, "I cannot employ these people, so I would rather see you and Prof. Scott have the chance than anyone else," meaning Marsh. Then in 1889, Cope obtained some relief from his trials when the University of Pennsylvania hired him as a professor of geology and mineralogy, ending what the writer David Rains Wallace called "the unseemly spectacle of a major American naturalist hung out to dry for a decade."[7]

Infinitely more unseemliness lay just ahead. Powell had previously offered Cope a salary and the opportunity to publish within the Hayden Survey, on condition that he donate a set of fossils to the National Museum of Natural History. But like so many other tantalizing prospects in Cope's life, this went nowhere. In the fall of 1889, still hoping to publish the second half of his bible, Cope wrote to demand that the USGS pay him for the years he had worked as a volunteer and refund the $75,000–$80,000 he had spent collecting fossils. Alternatively, it could pay him $3,800 a year for three years to work on getting his second volume ready for publication. On receiving this ultimatum, Powell snapped. Secretary of the Interior John W. Noble passed along to Cope Powell's decision that he would be happy to put an estimate before Congress for the remainder of the publishing Cope had in mind. But first "the government collections now in your hands at Philadelphia should be transferred to the National Museum at Washington."[8]

Cope was apoplectic. His wealth consisted of little other than his fossil collection. In his reply of December 20, he cried robbery, invoked the hated name of Marsh (who had supposedly delayed publication of Cope's second volume so that he could plagiarize it), and concluded, "The material in my possession belonging to the United States government will not exceed a bushel in quantity."[9]

In his fury, Cope turned loose the reporter to whom he had been feeding twenty years of accumulated Marshiana. During Christmas Week, the *New York Times* later reported, "a long, rambling statement of charges of plagiarism and incompetency . . . was hawked about among the newspaper offices." The *Times* deemed it "unworthy of consideration" (though it claimed so only after the story had made a splash in a rival paper). Cope's newspaperman was freelancer William H. Ballou, almost all of whose impressive scholarly credentials would turn out to be either undocumented or utterly fictitious. Historian Elizabeth Noble Shor, who methodically researched his résumé, dubbed him "Ballou, the Unbelievable One." Even Cope considered him a "rough customer."[10]

Readers of the *New York Herald* opened their paper on Sunday, January 12, 1890, to the headline "Scientists Wage Bitter Warfare." The story ran on for more than eight solid columns of type, interrupted only by thumbnail sketches of Cope, Marsh, and Powell, decorated with multiple sensational subheads: "Corroboration in Plenty," "Red Hot Denials Put Forth," "Will Congress Investigate?" "Can This Be True?"[11]

No longer working behind the scenes, Cope charged Marsh and Powell with scheming to elect Marsh president of the National Academy of Sciences so they could use its power to change the course of government science and put allies from the academy on the federal payroll. To preclude the possibility of competition, Marsh had supposedly sent USGS collectors into the field with instructions to sweep up the best fossils and "smash" everything else. Ballou also charged Marsh with "retaining" the salaries of his field parties in addition to his own $4,000 salary. "But the most astounding charge against Professor Marsh is that all of the work purporting to be his, as published by the government, is not his own, but in part that of his employés, the remainder being a collection of plagiarisms." Cope complained that he had been ordered to turn over collections gathered "at an expense of about $80,000 of my own money" to the National Museum. Meanwhile, he said, Marsh had spent $60,000 a year of government money on collections that were "all stored at Yale College, with no assured record as to what belongs to the government and what to the college."[12]

According to Ballou, a half dozen past or present Peabody staffers had supplied information for the article. He quoted a letter to Cope from Samuel W. Williston calling Marsh "a liar" and adding, "I never knew him to do two consecutive, honest days' work in science . . . when in New Haven he rarely appears at the museum till two o'clock or later and stays but an hour or two, devoting his time chiefly to the most absurd details and old maid crotchets." Marsh's publications were largely "the work or the actual language of his assistants. At least I can positively assert that papers have been published on Dinosaurs which were chiefly written by me."[13]

To avoid a libel charge, Ballou had circulated the article among the affected parties before publication. Under the subhead "Suppress It? Not Much!" he noted (and dismissed) the complaint from Princeton's W. B. Scott that he was an unwilling participant in the article. Ballou had come to meet him not as a journalist, Scott later claimed, but pretending to be an amateur scientist. Marsh was soon forwarding letters from his subordinates and from his enemies Scott and Henry F. Osborn asserting that they had not authorized the newspaper to use their names. Williston protested that he had written his letter to Cope in a moment of exasperation, probably following the early death, in 1887, of Peabody scientist Oscar Harger, who had contributed heavily to Marsh's work.[14]

Powell also issued a detailed and generally dispassionate refutation of the charges. But he stooped to sarcasm when it came to the charge that support from the National Academy of Sciences and participation of university scientists in the USGS somehow constituted a conspiracy: "And what is it, forsooth, that all these men have combined to accomplish that is so wicked? Nothing less than to put under a bushel the light of the glory of the genius of Professor Cope." He also praised Cope, if only to suggest how regrettable this public airing of grievances was to almost everyone: "The Professor himself has done much valuable work for science. He has made great collections in the field and has described these collections with skill. . . . If his infirmities of character could be corrected by advancing age, if he could be made to realize that the enemy which he sees forever haunting him as a ghost is himself . . . he could yet do great work for science."[15]

Cope returned to the pages of the *New York Herald* to assert, as if wrapped in the American flag, that he had acted only in the interest of the nation. "I should not have cared to expose Marsh's career had it not been for the injudicious zeal of his friends in elevating him to the highest place in the gift of American science," he averred. "My patriotism rebels against this gross error and, for one, I wish to wash my hands of it. I refuse utterly to have my criticism of Professor Marsh put on the low ground of personal quarrel."[16]

The sensation continued, at least in the pages of the *Herald,* to the middle of the week. Other papers paid little attention, and Marsh himself would surely have done better to stay above the fray, letting Powell's dignified reply stand for them both. But his vindictiveness was as unstoppable as Cope's perennial sense of injury, and his reply, when it came, was a bid to reach past the jugular and rip out Cope's soul. He began reasonably enough, asserting, as had Powell, that the paleontological work for the survey had cost more like $12,000 a year, not $60,000. Rather than pocketing his $4,000 salary, Marsh said, he applied it to the work. (He refrained from adding that he had also never received a salary from Yale.) He amply acknowledged the assistance of his staff, particularly Harger and even the apparently treacherous Williston and Hatcher. But he also said that every detail of his work "was performed by myself or directed by me." In any case, the idea of scientists, politicians, or even Supreme Court justices putting their names on work drafted by subordinates was hardly shocking.[17]

Marsh maintained (falsely, it seems) that there had been no mixing of the government collections with those of Yale, and that all were available for scientists "of good moral character." Then he joined Cope on the low road, illustrating bad moral character by recounting how Cope had made a surreptitious "raid" on the Peabody on a Saturday afternoon in 1882 "when the workrooms of the Museum were closed and most of the attendants absent." Benjamin Silliman Jr., who was working with Cope then on some misbegotten mining investment, had unwisely admitted him. Cope had wandered through Marsh's private rooms, "where the results of years of my labor were spread out ready for publication" and "where some of my rarest fossils were stored." Beyond the implied intellectual theft, Marsh then charged Cope with actual theft of specimens on a prior occasion and added that Cope had a reputation for light-fingered "depredations on the museums of the scientific world."[18]

To Cope's charge that Marsh had plagiarized his genealogy of the horse from the Russian paleontologist Vladimir Kowalevsky, Marsh replied, "My conclusions were based on specimens I collected myself. I never saw Kowalevsky's work until my own was completed and partly published." In passing he mentioned that Kowalevksy, like Cope, had a reputation as a specimen thief. Then Marsh made a remarkably callous comparison: "Kowalevsky was at last stricken with remorse and ended his unfortunate career by blowing out his own brains. Cope still lives, unrepentant."[19]

Finally, Marsh went on for six long paragraphs recounting Cope's embarrassing "Wrong End Foremost" blunder with *Elasmosaurus.* Marsh's assertion that he was the one who pointed out the error, and that Leidy merely took a supporting role, was probably false. It was Leidy who actually published Cope's error. So it also seems unlikely that this was the true start of Cope's lifelong animosity toward Marsh. But it was a convenient story to distract readers from his own bad behaviors while also humiliating Cope both for his error and for being motivated even twenty years later by "his wounded vanity." In describing Cope's frantic effort to retrieve all copies of the published work with the animal's head mistakenly positioned on its tail, Marsh also inadvertently revealed that he had been collecting Copeana almost as assiduously as Cope had been piling up his Marshiana. "I returned to Professor Cope at his request the one he sent me, but have two others, which I since purchased."[20]

This continuing spectacle of punch and counterpunch made Cope, Marsh, paleontology, the USGS, and American science at large a laughingstock. "All savantdom is agog," the *Chicago Tribune* reported with mixed feelings of horror and glee, at the spectacle of paleontologists and geologists "careering about as sharpshooters and outriding guerillas."[21] Satirists lampooned the quarrel in clumsy verse. The comedy was a disaster for the two participants, blighting their careers and forever obscuring some of the greatest achievements in the history of the biological sciences.

"Cope May Be Removed" was the headline in the *Philadelphia Inquirer* just two days after the first *New York Herald* story. Marsh had applied pressure on University of Pennsylvania president William Pepper to fire Cope. As the university trustees gathered to consider his fate, Cope in turn had urgently called Henry Fairfield Osborn in from New York to vouch for the truth of his accusations. He managed to hold onto his newly acquired position after being admonished about the need for scholarly decorum.[22] Though he was able to defend his fossil collection as his own, not the government's, he was eventually obliged to sell much of it at a loss to the American Museum of Natural History.

In 1894, Cope's wife, Annie, left him, "finding it too hard to live with a genius," according to Osborn, but also because of financial troubles and "many tales of ladies" in his past. He was thus largely alone in his crowded workrooms during the last months of his life. In February 1896, Cope was immobilized by an excruciating urogenital condition (he called it cystitis, a bladder inflammation) but continued to work on his fossils. He treated himself with formalin, a formaldehyde solution otherwise used for embalming, which left Osborn "somewhat aghast" and urging hospitalization. But Cope continued to self-medicate, writing to his wife soon after, "I took an enormous quantity of morphia and finally resorted to belladonna . . . the blockade was broken, and the agony ceased." On April 12, 1897, aged fifty-six, he died in his bed, "on all sides of which fossil bones were piled."[23]

There had been no rapprochement between the great Bone Wars rivals, and perhaps predictably Marsh made no public comment or gesture of respect on the death of his enemy. His own career had also turned downward in the aftermath of the Bone Wars debacle. The USGS had made enemies in Congress because of John Wesley Powell's insistence on planning the settlement of the West on scientific grounds, taking account of the availability of water. When opponents pushed to develop the entire West, Powell warned accurately that they risked "piling up a heritage of conflict and litigation over water rights."[24] Marsh became the unfortunate proxy for this fight.

In 1892, an Alabama congressman named Hilary Herbert, a conservative with no interest in science, held up a copy of Marsh's *Odontornithes* monograph to denounce the idea of spending taxpayer money on "atheistic rubbish." It was the perfect volume for his purposes, lavishly printed with gilt edges and Morocco leather binding. O. C. Marsh quickly wrote to explain that this was a special edition he had paid for himself, before he joined the USGS. But Herbert had no interest in facts. The thought that taxpayers had paid for a book about "birds with teeth"—*birds with teeth!*—became a rallying cry for the antiscience forces. Incredibly, the vertebrate paleontologist Henry Fairfield Osborn joined the attack on vertebrate paleontology (at least as long as Marsh was in charge). He argued that studying invertebrate fossils would be a more practical way to spend taxpayer money because they were a better tool for calculating stratigraphic dates.[25] In the end, Congress slashed the USGS budget by 20 percent, leading to Powell's resignation.

Marsh managed to hang on briefly at the USGS in an honorary role, but was then abruptly dismissed. He had spent his own fortune on fossils and on a lavish way of life, buying costly collections of orchids and Japanese art to adorn his grand estate. His uncle George's trust funds also ran out of cash in the recession of the early 1890s, and they were now, as a cousin put it, "like a squeezed orange."[26] Without his former government budget, he was forced to lay off the invaluable collector John Bell Hatcher and the Peabody's head preparator Adam Hermann. Princeton's W. B. Scott snapped up Hatcher (with funding from Osborn), and Osborn himself grabbed Hermann for the American Museum of Natural History. The engine of paleontological scholarship Marsh had been running since 1870 began to sputter. Marsh would eventually be obliged to request a salary from Yale and mortgage his home to the university for $30,000.

Osborn now turned out to be Marsh's most formidable enemy. He was too shrewd to repeat Cope's and Marsh's mistake, by draining his own fortune to build up his paleontology department. Instead, he cultivated donors, persuading them, according to biographer Ronald Rainger, by presenting fossils as "rare, valuable, and visible facts" for public display. Rather than lash out furiously at his enemies, as Cope had done, Osborn executed his subversive attacks with strategic precision. Marsh had for years used his position with the USGS to block Osborn from access to public and Indian lands. When he attempted to do so again in his newly weakened state, Osborn worked around him through influential friends at the federal level and also told Marsh, in effect, "Let's take it to the trustees of the Peabody Museum and see what they think." Yale had continued to back Marsh, but it must also have been embarrassed by revelations about his autocratic behavior. Marsh demurred. At every turn, "Osborn's aim was to undermine Marsh," wrote Rainger, "and he succeeded."[27]

Marsh continued to collect and publish in the 1890s, producing two influential monographs, *Vertebrate Fossils of the Denver Basin* and *Dinosaurs of North America*, both issued by the USGS. The latter solidified Marsh's contribution as developer of much of the modern language for classifying dinosaurs—that is, as Theropoda, Sauropoda, Ornithopoda, and Ceratopsia. On Christmas Day 1897, the *New York Times* reported that the French Academy of Science had honored Marsh with its Cuvier Prize, awarded only every three years for "the most remarkable work" in zoology or geology.[28]

The following year in New York, Marsh gave a talk to Yale alumni, recounting in a jocular tone some of his experiences in the "bone business." His great remaining ambition then, biographers Schuchert and LeVene wrote, was to complete the Peabody Museum building. His talk that night led up to the question of what to do with his enormous private collection of fossils, on which he had spent upwards of $250,000 and a lifetime of work. "I, of course, could not take them with me," he joked, "for if I did, they would probably burn up, as some of our clerical friends believe." Instead, he had decided to donate the entire collection (and though he did not say so, all but a fraction of his estate) to Yale and the Peabody. One reason for the gift, he said, was the hope that it might inspire "some generous friend of the University or of science" to "give the small sum necessary to complete the main museum building." He joked that Mother Yale sometimes struggled with fund-raising for theological reasons: it "was not quite good enough to secure the saints," who went to Princeton, "nor quite wicked enough to catch the sinners," who went—need one say?—to Harvard. Then he repeated his bid for a donor. No one stepped forward. In the last years of his life Marsh's "constant importunities to this end made him unwelcome in many quarters," Schuchert and LeVene wrote. The Peabody Museum remained unfinished.[29]

In late February 1899, Marsh made a trip to New York and Washington to negotiate the end of his work for the USGS and the return of those fossils his staff had gathered at government expense. (Many of them are even now among the most prized attractions of the National Museum of Natural History.) On the trip home, Marsh was caught in a heavy rain and developed a debilitating cold. He continued to come into the museum, with staff helping him up to his office via the freight elevator. He was treated by his personal physician, the same T. H. Russell who had turned up the nearly complete head of the toothed bird *Hesperornis regalis* on the 1872 student expedition.[30] But Marsh's condition worsened, confining him to his bed with pneumonia. On March 18, 1899, he died at home, aged sixty-seven. He was buried at Grove Street Cemetery in the heart of Yale, which had become his mother, and close by the Peabody's menagerie of ancient creatures, which were his beloved children.

CHAPTER 15

Trilobite Magic and Cycad Obsessions

For oh, it was a happy plight,
Of liberty and ease,
To be a simple Trilobite
In the Silurian seas!
MAY KENDALL, *Punch,* 1885

Among the seventeen thousand known species of trilobite are some named for each of the four Beatles, and others for Sid Vicious and Johnny Rotten of the Sex Pistols. Trilobites also have a Facebook page for their fans, and why not? Beginning about 500 million years ago, these many-legged little ocean dwellers, cousins to our horseshoe crabs, dominated the planet as thoroughly as dinosaurs did in their epoch, and as humans do now. They also flourished longer than either, hunting and scavenging and filter feeding across the seabeds of the world for 270 million years.

Many early human cultures attributed magical protective and curative powers to trilobite fossils, moved, as we are today, by something aesthetically satisfying, and a little mysterious, in the segmented carapace and the shieldlike shape. More than fifteen thousand years ago in central France, Cro-Magnon people drilled holes in a trilobite fossil, presumably to hang it from a string as an amulet. The Ute Indians of southeastern Utah also wore them on necklaces, and sometimes buried a loved one with a trilobite resting on the breastbone. They gave trilobites a name that, loosely translated, means "little water bugs in stone houses."[1]

Beyond their supposed magic, no one knew much about these ubiquitous fossils until the late nineteenth century, to the great frustration of paleontologists. They could find trilobites almost everywhere, but none of them showed any ana-

In the era before X-rays, Charles E. Beecher used meticulous handwork to reveal the appendages and inner anatomy of this fossilized trilobite, now prized by Peabody staff as catalogue number YPM 228.

tomical features of the underside, presumably because the softer organs and appendages there had dropped off and decayed, while the harder carapace became fossilized. It wasn't until the late 1870s that scientists accepted for the first time that trilobites had legs, when a young geologist named Charles Walcott began to write about the specimens he had excavated on his in-laws' farm in upstate New York. But Walcott soon moved on to other work. (He would in time succeed John Wesley Powell as director of the U.S. Geological Survey, discover one of the world's great fossil fields, the Burgess Shale in the Canadian Rockies, and finally become secretary of the Smithsonian Institution.)

At that point, improbably, a one-armed carpenter living in Rome, New York, thirty miles down the road from Walcott's discoveries, took up the hunt. For eight years, amateur paleontologist William Seymour Valiant and his half brother Sidney Mitchell spent their day off each week excavating in the shade of some trees at a place called Cleveland's Glen. Finally, their work yielded the prize they were seeking—a trilobite with what appeared to be antennae, legs, and "swimming appendages." But "the discovery of feet and other organs of the trilobite," Valiant later recalled, had been announced so often before "based on insufficient evidence, or no evidence at all, that naturalists were disinclined to accept any statement that such discoveries had been made."[2] Finally, in May 1893, a Columbia University professor read a paper about the find before the New York Academy of Science.

Peabody paleontologist Charles E. Beecher was soon en route to Rome. Even in that year's unexpectedly straitened circumstances in the paleontology department, this was an irresistible collecting opportunity for the museum—if not quite in its own backyard, then in the next-door neighbor's. That June, with permission from Valiant and the landowner, Beecher extracted almost a ton of rock that included the narrow band known as "the *Triarthrus* zone," which in places was just a quarter-inch thick. Beecher was a modest figure, described by colleagues as kind, "quiet and unassuming," with no interest in adulation—that is, the polar opposite of O. C. Marsh. But in a move that had Marsh's fingerprints all over it, Beecher paid $25 to purchase an agreement from the landowner giving the Peabody the right to collect fossils there for a period of ten years, excluding everyone else except Valiant and his brother. "I feel quite well satisfied now with the results of this trip," Beecher wrote to Marsh, sounding like a speculator who had just cornered a precious commodity market, "and think we can nearly control the antennae business. I look forward with pleasure to working up the collection."[3]

The excavation area would later become known as "Beecher's Trilobite Bed," and two factors made it special. The first had to do with the nature of the geology,

Beecher (*seated, right*) in upstate New York, where he found exquisitely preserved trilobites.

which had replaced the trilobites' living tissue with pyrite, or fool's gold, outlining every detail of their appendages against the black shale substrate. The second factor was Beecher himself, who was an artist at the tedious business of picking away stone, almost atom by atom, to free—and not destroy—the fossil hidden within.

He started with hammer and chisel, then moved on to dental chisels, sometimes turning the specimen on a dental lathe. Beyond that, the finer work seemed almost too fine for human hands. As in many crustaceans today, the trilobites' limbs forked and were fringed with filamentous beards. In death, limbs also lay one atop another, higgledy-piggledy, in specimens that sometimes measured less than an inch in length. But "aided by his remarkable manual dexterity, mechanical skill, and untiring patience," the Smithsonian's William H. Dall wrote, Beecher "worked out the structure of antennae, legs, and other ventral appendages with a minuteness which had previously been impossible."[4]

At first, Beecher experimented with chemical removal of the surrounding rock, but this dissolved the trilobite together with the shale. He switched to carborundum, pumice, and other abrasives, which had the advantage of removing the soft surrounding shale more quickly than the fossil within. After thousands of trials, he developed his method of rubbing an abrasive into a specimen with a small rubber pad, which he then made smaller and smaller to fit in the infinitesimal spaces between the anatomical parts his work was revealing. In some cases, Beecher then took the completed side of a specimen and turned it over, embedding it in Canada balsam on a microscope slide so he could begin to work in from the opposite side. He did this with *Cryptolithus* trilobites, which were rare, and according to Charles Schuchert, who witnessed the work, two such specimens "revealed both sides of the individuals, though they were then hardly thicker than writing paper." Because his methods were unknown, and because he published his trilobite anatomy at first only in the form of drawings, some scientists regarded his results, Schuchert added, "as startling, as iconoclastic, and even unreliable."[5] Photographs proved more persuasive.

Beecher went on to prepare many more specimens in this fashion, paying particular attention to the inch-long *Triarthrus eatoni.* His detailed drawings and descriptions helped to define it as both a scavenger of organic debris and a predator on worms and other soft seabed-dwelling creatures. "Using its spiny legs," the modern trilobite researchers Derek E. G. Briggs and Gregory D. Edgecombe have written, "it passed these morsels along its belly from back to front toward the mouth before devouring them, much as a horseshoe crab does today."[6] Beecher had rescued trilobites from the realm of magic and revealed them as real animals.

"Mostly a Sunday Matter"

Among the few other scientists still working in paleontology at the Peabody in the period just after Marsh's death was a doctoral candidate named George R. Wieland. In 1895, before he was affiliated with the Peabody or anywhere else, he had made a spectacular discovery on a solo expedition near the Black Hills in South Dakota. It was a monstrous turtle found embedded in the side of a ravine on the south fork of the Cheyenne River. He used it as "my *card of introduction*" on applying for a fellowship at the Peabody in 1898.[7] Marsh promptly awarded him the fellowship and purchased that turtle and a second specimen for $200.

Like many Marsh acquisitions, *Archelon ischyros,* as Wieland had named it, would remain in its crate for years afterward. But when it emerged in the first decade of the twentieth century, it became one of the most popular displays in the museum's history. The head was missing, and Richard Swann Lull, a new staff paleontologist, manufactured a model of it based on another, smaller skull Wieland had collected. The Peabody Museum preparator Hugh Gibb pieced the parts together, and the two of them debated whether to replace the missing right hind foot. But the remaining bones had healed, and the thigh bone on that side was shorter, apparently from disuse, meaning an old injury, as if it had been bitten by a mosasaur. "The animal was left mutilated," Lull wrote.[8] The finished turtle measured eleven feet from tail to snout, and Gibb stood it on end to emphasize its scale.

George R. Wieland had discovered the giant fossil sea turtle *Archelon* and made it his "card of introduction" when seeking a museum fellowship under Marsh.

Wieland intended to continue working with vertebrates, and in 1898 Marsh sent him back to South Dakota to collect more turtles and to excavate a giant dinosaur, *Barosaurus,* which he had asked a local family to hold for him. But he also asked Wieland to visit a nearby ridge that was strewn with a forest of fossilized plants called cycadeoids, also known as cycads for their resemblance to a variety of modern plant with a woody stem and palm-like crown. Marsh wanted Wieland to settle a dispute about whether this petrified forest was of Jurassic or Cretaceous origin. Wieland declared them Cretaceous: How another researcher "ever figured them to be anywhere near Jurassic is a mystery to me." He also assured Marsh that he wasn't taking too much time away from *Barosaurus* for this side project: "The Cycads have been mostly a Sunday matter," he wrote.[9]

The truth, though, was that Wieland was now hooked on cycads. They had obvious visual interest, with their intriguing shapes, often resembling a beekeeper's traditional basket beehive, and with a surface pattern of orderly pockmarks from old leaf attachment points. Local ranch families who collected the fossilized trunks as curios described them, Wieland wrote, as "beehives," "wasps' nests," "corals," "mushrooms," and even "beefmaws" for their resemblance to a bovine reticulum, the first digestive organ of a ruminant's alimentary tract.[10]

Wieland, who was rapidly morphing from vertebrate paleontologist to paleobotanist, desperately wanted to examine the interior surfaces to see what they might reveal about plant evolution. Back at the Peabody Museum, he tried working on them with chisels and with a drill powered by an air compressor. Then he used tube-shaped drill bits to extract core samples, first boxing up a specimen in wood and securing it at the desired angle by embedding it in plaster. The task required "skill, time, knowledge of the general structures, and constant watchfulness," Wieland wrote, "lest the cylindrical core containing the fructification be broken away and shattered into pieces before the drill can be stopped." He began making cross-section slices of cycads using a rock-cutting machine built by Benjamin Silliman Sr. and run by foot pedal. In 1900, the museum trustees approved the purchase of an engine for running a more sophisticated rock-slicing machine.[11] This enabled Wieland to cut his cross-sections at any angle through the rock. They turned out, when carefully polished, to be warmer and more beautiful than marble, with soft brown patterns indicating the shapes of ancient seeds, cones, and other structures. To get multiple samples from the same rock, Wieland would make a mold to preserve the original dimensions, secure the specimen in plaster, and saw out his vertical slice. Next he glued the two slightly reduced halves back together with a stone cement section in between to compensate for the lost cut and make a new cut horizontally. These techniques, borrowed from local gravestone cutters, enabled him to examine cycad anatomy in extraordinary detail.

On repeated trips back to the site in South Dakota, Wieland retrieved more than seven hundred cycad specimens, giving the Peabody the most extensive collection of these fossil plants in the world. But he also became alarmed that collectors with no scientific interest were carting off cycads purely as curios. In 1920 he filed a claim under the Homestead Act to the 320-acre site containing what was left of the ancient forest. Then he relentlessly badgered federal officials to take over the site and protect it "in order that the cycads might not fall into unworthy hands."[12] Two years later, President Warren G. Harding designated it the Fossil Cycad National Monument. But protection was sporadic at best.

Wieland would later become caught up in a feud with a federal geologist named Carrol Wegemann, who blamed him for having taken all the cycads back to Yale before donating the land to the government. Wegemann also accused Wieland of stealing fossils during a 1935 excavation, a charge that is still sometimes repeated in South Dakota today. (The Peabody says it holds no fossils from the site collected after 1916.) Wieland continued to lobby for a field museum at the site, even staging a design competition among Yale art students. But he also acknowledged that there was nothing left at the surface to gratify "the untrained

By meticulously cutting thin slices of fossil plants called cycadeoids, Wieland could examine their internal anatomy in detail.

perception of petrified forests."[13] Many people began to question the value of the national monument designation, and it was finally withdrawn in 1957. A highway right-of-way later cut through the site, revealing additional cycads.

The Crown Prince and the Paleobotanist

Over the years, Wieland became obsessed with his subject, even by the standards of museum curators. He could talk about nothing else, and he seems to have talked endlessly, to the point that it became necessary to ban him from the preparators' workrooms. In time, he "was practically kicked out of the museum," according to an oral history, and had to move "his big old saw set up" to the basement of another building on campus. "When did he begin to sort of go to pieces? Was that before he retired or after?" one former director of the museum wondered in that oral history, and another former director replied, "Maybe it was when he was born. He was a queer one all his life."[14]

In June 1926, Yale was preparing for a visit by Sweden's crown prince Gustaf Adolf VI, who was receiving an honorary degree. University authorities were "agog over the rare opportunity of entertaining royalty," a Yale alumnus later recounted in a letter to *Time* magazine. They prepared to mark the event "with the utmost dignity and propriety," beginning with "an exclusive little luncheon for only the mightiest figures of academic and of local society." Then the prince arrived and asked for the one thing no one had ever thought to include: George R. Wieland. "Consternation smote the party and a frantic search for Yale's forgotten man ensued."[15]

Wieland, who had no phone, was eventually tracked down at his suburban home, hustled to New Haven, and seated beside the prince. "The conversation presented pretty tough going for the local elite and even for the President and Fellows, for it dealt almost exclusively with fossil cycads." The others were praying for the luncheon to end so they could escape the stupefying talk of cycads and perhaps otherwise engage their honored guest's attention. Instead, the prince suggested a visit to Wieland's workroom. Someone raced ahead in a desperate bid to bring order "out of the monumental chaos and dustiness" there, to no avail. The dignitaries found it "a little difficult to appear dignified and interested and at the same time keep their morning coats and striped trousers out of the inch-thick dust while the Prince and Wieland continued their ardent and interminable conversation."[16] The prince, it was evident, was a serious paleobotanist, and George R. Wieland had finally found his perfect audience.

Bringing Dinosaurs to Life

Marsh's death had come just four years after James Dwight Dana's, and hardly anyone would have been a natural heir to the public stature of two such giants. Nor had Marsh's leadership style anticipated the unthinkable possibility that he might ever need an administrative or intellectual heir. Thus Charles E. Beecher became head of the Peabody's geology division, though he seemed ambitious for nothing more than time to free trilobites from their stone prisons. But Beecher almost immediately undertook a bold initiative having nothing to do with trilobites that overturned, in the most public fashion, one of Marsh's most beloved policies.

Twenty-five years earlier, the Smithsonian's Spencer F. Baird had written Marsh that he was under pressure to mount "some restorations of prehistoric animals" for the celebration of the American centennial in 1876 in Philadelphia. Waterhouse Hawkins had created a plaster model of a duck-billed dinosaur at the Academy of Natural Sciences in 1868. But Baird didn't really trust Hawkins or think making a mount was a wise use of real fossils, and Marsh agreed: "In the few cases where the materials exist for a restoration of the skeleton alone, these materials have not yet been worked out with sufficient care to make such a restoration perfectly satisfactory." He was apparently worried about presenting spectators with an unforgettably huge mistake—especially in Cope's hometown. "A few years hence we shall certainly have the material for some good restorations of our wonderful extinct animals," he added, "but the time is not yet."[17]

It would not come in his lifetime. "Marsh saw the desirability of making these strange beasts come to life in authentic skeletal restorations that would visualize their former appearance," wrote Schuchert and LeVene.[18] But he did so only on paper. He started by having his artists make a detailed drawing of each of the pieces of a skeleton, to be cut out so he could move them around and figure out how each part articulated with the others and how each should be positioned. Only when he had satisfied himself in this fashion would he have a complete restoration drawn and committed to stone by a lithographer. Marsh became a pioneer in publishing restorations in this fashion. He also publicized his finds throughout the scientific world by making casts of bones for other museums. But the bones themselves mostly remained carefully cushioned and locked away in the storerooms of the Peabody, to be contemplated only by the sacred priests of paleontology.

Within months of Marsh's death, Beecher set out to change this. The museum had a nearly complete specimen of a duck-billed dinosaur, discovered eight years earlier by John B. Hatcher in Wyoming. Marsh had named it *Claosaurus* (now

In 1901, this *Claosaurus* (now *Edmontosaurus*) was the first mount of the actual bones of a dinosaur in North America. But many paleontologists disliked the dynamic, modern-looking pose.

Edmontosaurus) *annectens* and stored it away as usual. Many of the bones were embedded in slabs of Laramie sandstone in their original positions, and Beecher set out with preparator Hugh Gibb to combine the slabs to create a huge wall-mounted plaque. Raised back up on its feet, the dinosaur would stand more than fourteen feet high and twenty-six long. (It still stands a century later, on the north wall of the museum's Great Hall.)

Beecher's idea wasn't just to reveal the dinosaur as in life. He and Gibb wanted to convey the impression "of the rapid rush of a Mesozoic brute." Beecher's own description conveyed some of the intended excitement: "The head is thrown up and turned outward. The jaws are slightly separated. The fore arms are balancing the sway of the shoulders. The left hind leg is at the end of the forward stride and bears the entire weight of the animal. The right foot has completed a step and has just left the ground preparatory to the forward swing. The ponderous and powerful tail is lifted free and doubly curved so as to balance the weight and compensate for the swaying of the body and legs. The whole expression is one of action and the spectator with little effort may endow this creature with many of its living attributes."[19]

The effect was spectacular enough to make the front page of the *New York Times* in April 1901. It was the first mount of the actual bones of a dinosaur in North America, and only the second in the world, after an 1883 mount of an iguanodon at the Royal Belgian Institute of Natural Sciences in Brussels. (The *Times* thought "the American monster" clearly superior.) It was also the beginning of a rush by museums to bring dinosaurs out of the storerooms and thrill the public by presenting them as they had lived. But the unfinished Peabody Museum itself had too little room then to display its many other dinosaur treasures, as the *Yale Alumni Weekly* later wryly noted:

Methinks I hear in chorus
Each half-mounted Brontosaurus,
Each Iguanodon, Pteranodon, and Spoon-Bill Dinosaur,
Cry against their profanation:
"O respect our age cretacian,
Give us room to live our lives out! Can't you set us up once more?"[20]

Beecher was otherwise not inclined to take on Marsh's role as the lion of American paleontology. In any case, Henry Fairfield Osborn of the American Museum of Natural History was eager to do so, replacing Marsh in 1900 as vertebrate paleontologist for the U.S. Geological Survey. Despite Beecher's reluctance, Osborn was able to negotiate access to Marsh's fossil collections at the Peabody. He sought, according to Osborn biographer Ronald Rainger, "to establish a normalized working relationship with Yale," which was of course long overdue.[21] But that gave Adam Hermann, the preparator Osborn had hired away from the Peabody, the information he needed to create the first mount of *Brontosaurus* decades before the Peabody was able to mount its own prize specimen. Osborn also successfully lobbied Charles Walcott, the survey's director, to let him take over Marsh's unfinished monographs, though Beecher had hoped to keep them at the Peabody. Happily, Osborn delegated the monograph on *Triceratops* and its ceratopsian kin to John B. Hatcher, who had almost single-handedly discovered them. But he kept Marsh's brontotheres for himself and sought to erase that name in favor of his own "titanotheres." It was as if a new alpha male had taken over the lion pride, promptly killing his predecessor's cubs. It was also the sort of aggressive behavior Marsh himself might have displayed.

"Osborn maintained that his efforts were responsible for establishing good working relationships among paleontologists, and to some extent his claim was accurate," Rainger wrote. "Yet despite Osborn's rhetoric, such efforts served primarily to advance him and his department" at the American Museum. Like Marsh,

Osborn "gained control over important financial and institutional resources for doing work in the subject and employed the power associated with those positions to establish a dominant position for himself and his department."[22] Thus the American Museum took over from the Peabody as the leader of paleontological discovery at the start of the new century. It would be more than sixty years before Yale paleontology would rise again, as the instigator of the modern dinosaur revolution.

Beecher retired to his trilobites, and on a Sunday in February 1904, as he was working at home on a drawing of *Cryptolithus,* he suffered a sudden heart attack and died, aged forty-seven.

CHAPTER 16

Mapping Ancient Worlds

Evidently you paleontologists are about as friendly as ever.

A geologist on the 1921 campaign against Charles Schuchert

The Peabody building James Dwight Dana had said "ought to be made to last at least 1000 years" was already under threat after just thirty. It was partly Dana's own fault. He had persuasively argued for making Yale "the first university in the leading nation of the globe." Yale had taken him up on it and was now rapidly expanding in pursuit of that ideal. It made administrators hungry for precious real estate around the Old Campus.

The beginning of the end came on Friday, May 26, 1905, when the Peabody's executive board voted to meet with Yale president Arthur Twining Hadley about soliciting a donation from the philanthropist Andrew Carnegie to enlarge the Peabody on the original plan. Hadley was an old Carnegie ally, having helped him found the Carnegie Institution in Washington, D.C. But it was otherwise an odd idea, given that Carnegie had launched his own natural history museum in 1895. The Carnegie Museum had also just mounted its celebrated *Diplodocus* and named it *carnegii,* presumably securing its hold on his loyalty.

Word came back from Hadley almost instantly that the Yale Corporation had a very different fate in mind for the museum. On May 20, the weekend before the Peabody trustees contacted Hadley, three Yale alumni had arranged to purchase the thirty-three-acre Hillhouse family estate at the top of Hillhouse Avenue, with the aim of passing it on to the university, and ultimately shifting the center of the campus uphill. "If we can put our museums and special schools on the Hillhouse estate," Hadley wrote in his annual report for that year, "we are free to use our land in the city for dormitories and recitation rooms."[1]

A few of Schuchert's beloved brachiopod fossils suggest the diversity of the fossil record.

The Peabody's executive committee acquiesced with blinding speed to "the possible transfer of the Peabody Museum to the Hillhouse property." The museum desperately needed more space, and the committee's June 8 vote asserted only one condition, that "the University is prepared to furnish an adequate amount of land and a sum sufficient to replace the cost of the existing building as it stands on the books of the Museum Trustees." On June 12, the Yale Corporation voted to confirm the agreement, and on June 20, Hadley was writing to Andrew Carnegie seeking a donation.[2] The shortage of money to make such a move or otherwise develop the Hillhouse property would be a continuing problem for Yale and the Peabody. But the deal was done, and the Peabody's position at the geographical center of the university was now effectively over. For another dozen years, as the search for funds continued, the Peabody's original half-finished building would persist, as if on life support. Then it would be torn down on short notice, with no new home anywhere in sight.

The Petrifaction of a Nanny Goat's Horn

About the time George Peabody was first establishing the museum that would bear his name, in 1866, an eight-year-old in Cincinnati was standing around watching a gang of men dig a trench near his home. After a while, the boy later recalled, "one of the Irish laborers threw a bit of rock up to me with 'Here, Johnny, here's something for you.'" The boy pocketed the rock and kept it for years after, first labeling it a "what is it" because no one could identify it. After reading a book by the pioneering naturalist Alexander von Humboldt, his father, an immigrant German furniture maker, suggested that it might be a *vertsteinerung*, a petrifaction, a fossil. Thereafter, the boy named it "the petrifaction of a nanny goat's horn" because that's what it looked like to him.[3]

The prospects that Charles Schuchert's taxonomy would advance much beyond that point were dim. His father's small furniture shop could barely support the family of six children. High school was not an option, much less college. But Schuchert kept up his geological interest, spending his free time fossilizing around Cincinnati. When he came home in shoes caked "with the sacred soil in which the fossils were entombed," his mother rolled her eyes: "You bring more dirt into the house than your rocks are worth." His father took him at the age of eleven to see a local geological collection, "which opened to me an unknown world," not least because of the startling news that Cincinnati, and his entire world, had once been at the bottom of a vast inland sea.[4] Schuchert was soon able to identify the nanny goat's horn as a fossilized coral. Then more immediate realities set in: at thirteen, he left school to join his father in the furniture trade. After a fire at the shop, his father suffered a breakdown, leaving the family dependent on Charles, the eldest child. Despite his increased responsibility, Schuchert still found time to develop a special interest in marine creatures called brachiopods, which were turning up everywhere in the local quarries.

More than thirteen thousand brachiopod species are currently known. They appear in the fossil record as far back as the early Cambrian period, more than 500 million years ago, and though far less abundant now, they continue to live in the oceans today. The longevity of the group makes trilobites, with their 270-million-year history, seem like a passing fancy. Researchers theorize that brachiopods evolved from sedentary, tubelike animals. These simple creatures extracted calcium phosphate from the sea and secreted it as a protective covering of overlapping plates. In brachiopods, those plates eventually became an upper and lower shell, hinged at the rear. They can thus look a lot like mussels, scallops, and other familiar bivalve mollusks. But brachiopods and bivalves are only distantly related, belonging to separate taxonomic phyla. Among other differences, the two shells in a brachiopod do not necessarily mirror each other with clamlike symmetry. Many brachiopods also attach themselves to the sea bottom with a fleshy stalk that protrudes through a hole in the shell and can lift them up out of the muck for more efficient feeding. And whereas bivalves draw water into the shell and filter out food particles with their gills, brachiopods have a flowerlike crown of tentacles called a lophophore. The tentacles and their cilia undulate to create a current and capture the tiny food particles that drift past as a result.

Brachiopods interested Schuchert partly because they have been around so long and have evolved into so many forms with the passing ages. They are also ubiquitous, sometimes having piled up in dense colonies, as mussels do, forming ridges on the seafloor. That combination of longevity, species variety, and abundance made them useful as an index species for determining the age of the surrounding rock. (More recently, scientists have learned to use the chemistry of the shells to gauge the temperature of the seawater in which they lived, making brachiopods a way to measure changes in climate.)

Verbal Daggers and the Shotgun

While making himself an expert on brachiopods, Schuchert also took classes to learn the art of lithographic illustration. This enabled him to find work describing and illustrating fossils with Edward O. Ulrich, curator at the Cincinnati Museum of Natural History. Ulrich trained him as a professional paleontologist and the two became close friends. It would be another of those early friendships that later devolve into bitterness and treachery. The two made a swap, with Schuchert taking all the brachiopods in Ulrich's collection, and Ulrich taking all of Schuchert's bryozoans, starfish, ammonites, and so on. It was the first of many such exchanges Schuchert would make with collectors around the world.

The resulting brachiopod collection caught the attention of one of the great geologists of the day. James Hall ran a highly productive geological laboratory—half training ground, half gulag for aspiring geologists—in Albany, New York. Hall's standard technique for fossil hunting "was to flatter and invite collectors to work with him in Albany and to bring their collections," wrote geologist Robert H. Dott Jr. "Commonly, when the apprentice moved on, however, his collection did not." Hall may have hastened the departure of apprentices by his penchant for "throwing vituperative verbal daggers" and for menacing the wretched object of his fury with "a stout cane or even a shotgun kept at the ready near his desk."[5] Schuchert lasted two and a half years there, departing finally in anger over Hall's plan to publish Schuchert's synthesis of North American fossil brachiopods under his own name. But he had learned a great deal from Hall and made some of the most important professional contacts of his life, and when he left, he took his brachiopods with him.

Schuchert went on to a series of paleontological odd jobs, first for the state of Minnesota and then with the U.S. Geological Survey. For a nine-month stint at the Peabody Museum in the early 1890s, he worked with his friend Charles Beecher, also a former Albany inmate, to help prepare beautifully complete crinoid fossils, embedded in siltstone slabs, for display at the Columbian Exposition of 1893 in Chicago. Then he went to Washington, D.C., to work with his old Cincinnati friend Edward O. Ulrich, now at the U.S. National Museum. Summers he spent in paleontological field work. From an expedition to Greenland in 1897 with polar explorer Robert Peary, Schuchert came away with a permanently deformed left hand, the result of field surgery by a local ship captain.

He was now "married to the science," as he put it.[6] His single-minded focus showed up, beginning in the 1890s, in his abundant production of scholarly papers, covering vast swaths of time and geography: "Cretaceous Series of the West Coast of Greenland," "On Upper Devonian Fish Remains from Colorado," "On the Faunal Provinces of the Middle Devonic of America and the Devonic Coral Sub-provinces of Russia," and so on. In 1904, when Yale was seeking a

Schuchert carefully recorded his acquisitions, beginning with this specimen in the museum's first invertebrate paleontology ledger.

replacement after Charles Beecher's sudden death, Schuchert stood out as the obvious choice, despite his lack of even a high school diploma.

Peabody lore holds that he and Ulrich had begun to quarrel partly because Ulrich married the only woman Schuchert ever loved.[7] But now Ulrich watched with envy as his former apprentice and friend, who had been a furniture builder until he was almost thirty, became Yale's professor of paleontology and historical geography, and curator of the Peabody's geological collections. Schuchert himself was feeling less enviable and more sick to his stomach. At age forty-six, never having taught a college class, he stood before a roomful of expectant Yale undergraduates and discovered that he was clueless about how to explain the endless, shifting sprawl of geological history.

As part of what he called the "somewhat painful metamorphosis from curator to professor," Schuchert began to make stratigraphic drawings on a blackboard and then "paleogeographic" maps depicting ancient land and seascapes of different periods. "The results proved at once that this mode of presentation is strikingly helpful," geologist Adolph Knopf wrote. It wasn't a completely new idea. James Dwight Dana had introduced maps of ancient geography in 1863. But Schuchert's maps depicted much narrower slices of time with far more detail. Moreover, "it became his absorbing task during the remainder of his life," Knopf wrote, "to add to these maps day by day all new information as it came along." A graduate student recalled the maps piled up on the drafting table, "worn and dog-eared from constant use, and from many pencilings and erasures."[8]

Schuchert published his first collection of maps in a 1910 paper, "Paleogeography of North America," depicting North America in fifty separate stages from roughly 550 million to 5 million years ago (that is, the early Cambrian to the Pliocene). *Science* described it in a lengthy review as "a storehouse of information" and "of first importance to the student of historical geology." The reviewer criticized "the lack of evidence on debatable points throughout the paper" and quarreled with many details. But he added that Schuchert had acknowledged "the imperfection" of these maps and was seeking the help of others "in making the maps agree with the progress of future discoveries."[9]

Overnight, Schuchert became the leading figure in the nascent field of paleogeography. He was soon also collaborating with his Peabody colleagues to produce the popular textbooks *Historical Geology* and a companion book, *Outlines of Historical Geology.* They helped to establish paleogeography as part of the standard curriculum in the field—and also made him moderately rich. Ulrich, who said he had first proposed such maps to Schuchert in the 1890s, began to brood on the supposed theft of his idea.

Paleontological Combat (Again)

At the Peabody, Schuchert worked to make the collections available to all scholars, even if that meant Henry Fairfield Osborn and his staff. "What I'm trying to do is make friends between all of us," he explained in 1906, "and we have at Yale much to live down." In that same spirit, he made Marsh's unpublished work on plesiosaurs available to Samuel W. Williston, one of the Peabody dissidents in the 1890 Bone Wars rebellion, who had gone on to become a professor at the University of Kansas. (Yale would later go a step further toward healing old wounds, awarding Williston an honorary doctorate.) Schuchert also tried to avoid antagonizing colleagues in his textbooks, going out of his way to find room for a grand geological system proposed by Ulrich, though there was little evidence to support or even make sense of it. For Ulrich, Schuchert's reluctance to promote his scheme more zealously was a betrayal, especially as the scheme failed to gain backing elsewhere in the geological world.[10]

A standard theory about the quarrelsome lives of scholars holds that academic

Missing the Drift

On Christmas Day 1910, the same year Charles Schuchert was publishing his first paleogeographic maps, a young German researcher browsing through a friend's new atlas became intrigued by the matchup between the Atlantic coasts of Brazil and West Africa.[a] The two continents echoed each other like a couple sleeping in the spoon position, though separated now by an ocean. Alfred Wegener, a junior faculty member at a German university, was no paleogeographer. His specialties were meteorology and astronomy. But he was also an Arctic explorer and a balloonist with a world record for endurance flight, and he was blithely fearless about raising academic hackles in pursuit of the apparent movement of the continents.

Wegener was soon compiling evidence of plants and animals that were often strikingly similar on opposite sides of oceans: the marsupials in Australia and South America, for instance, looked alike, and so did the flatworms that parasitized them. In similar fashion, the stratigraphy often dropped off on one side of the ocean only to pick up again on the other. It was as if someone had torn a newspaper sheet in two, and yet you could still read a sentence across the tear.[b] He cut out maps of the continents and arranged them together on a globe, like jigsaw puzzle pieces, sometimes extending borders to show how they might have looked before the landscape crumpled up into mountain ridges. The result was a single supercontinent, later named Pangaea, with a southern half that would become Gondwana.

In a 1912 lecture and a subsequent article, Wegener called the resulting theory "continental displacement," though it soon became better known as continental drift. Injuries he sustained in combat in World War I gave him hospital time to extend his idea into a book, *The Origin of Continents and Oceans*. An English translation appeared in 1924, but Wegener's idea was already stirring up bloody international debate, with Schuchert among the leading opponents. "Continents are not like corks on oceanic currents flowing around to suit the fertility of German imagination!' he declared in 1922. (He was apparently still insecure about his own family background after America's intense anti-German sentiments during the war.)[c]

Having risen to become dean of American paleontologists and a leading proponent of paleogeography, Schuchert was also sensitive to the intrusions of a newcomer on his turf (his own lack of educational credentials momentarily forgotten). "Facts are facts, and it is from the facts that we make our generalizations . . . and it is wrong for a stranger to the facts he handles to generalize from them to other generalizations."[d]

Schuchert was hardly blind to geological change. It was the substance of the many

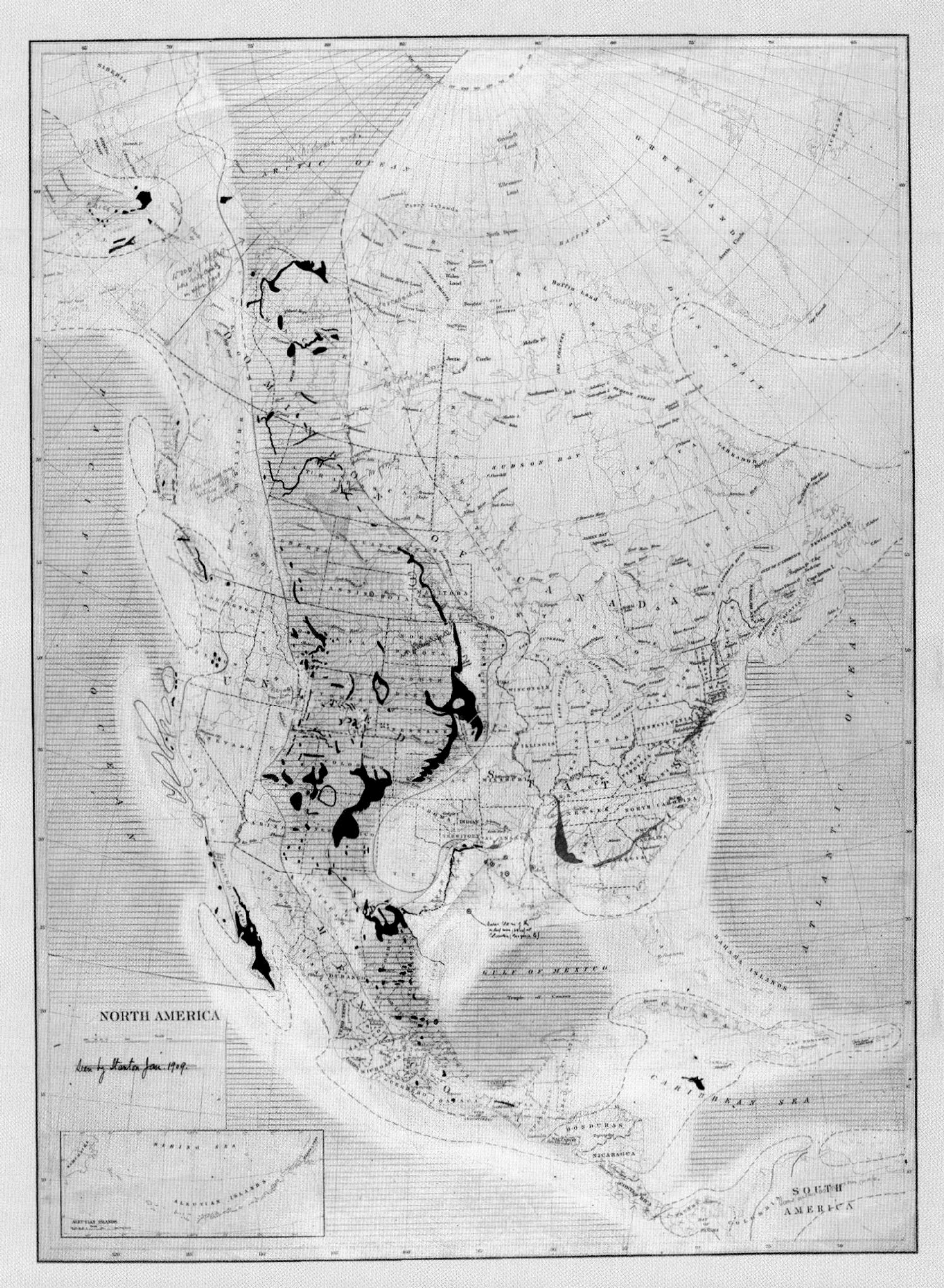

Schuchert continually updated and annotated his maps as knowledge about the history of the continents advanced.

paleogeographic maps he had made showing interior oceans and rising mountain ranges in ancient North America. He had also helped to establish many of the correlations between species on opposite coasts, deep into the past. But he maintained that the continents themselves had always held their present positions—according to the geological school of thought called uniformitarianism—and he explained those connections between opposite coasts with land bridges that later subsided into the oceans. This theory "had almost nothing in the way of supporting facts and raised many unanswered questions," according to geologist Roger M. McCoy. But Schuchert believed that geophysicists would eventually provide proof of the subsidence idea.[e]

He was operating less from facts and more from ideology, perhaps combined with institutional loyalty to the Peabody Museum: "The battle over the theory of the permanency of the earth's greater features introduced by James Dana has been fought and won in America long ago," he declared at a 1926 symposium on drift.[f]

Schuchert also rejected continental drift out of more practical paleogeographic concerns. Uniformitarianism provided a reassuring background of fixed continents and oceans against which to record the finer details of shifting geological history. "For Schuchert, abandoning uniformitarianism was nearly tantamount to abandoning historical geology altogether," wrote science historian Naomi Oreskes. "Not surprisingly, he declined to do both." Instead, he embraced a land bridge alternative that "influenced a generation of geologists to believe that drift was not so much impossible as unnecessary."[g]

Schuchert and his Peabody successor Carl Dunbar passed on their opposition to drift through their own students, among them George Gaylord Simpson, who remained a prominent antidrift holdout into the early 1970s. They also promoted the argument via the textbook *Historical Geology*, declaring in italics: "*There never has been general interchange in position between the continental masses and the basins of the oceans.*"[h] That textbook remained popular in college classrooms into the 1960s, when younger scientists finally broke down this old orthodoxy with irrefutable evidence that tectonic plates could literally move continents.

The paleogeographic maps that Schuchert helped pioneer live on even so. Paleontologists use them to identify formations that might be rich in fossils. Petroleum exploration crews study them for black shale deposits likely to yield oil, and mining companies count on these maps to lead them to subduction zones that might hold copper and gold. Scientists now also use them, as would Schuchert himself were he living, to record the past movements of tectonic plates and of the continents they push across the surface of the Earth.

rivalries are so bitter because the stakes are so small. But we all deal with pretty small stakes in our lives and have roughly the same experience of unforgivable slights and insufficient adulation. Maybe academic life merely gives its verbally inclined thinkers the freedom to brood about it for too long, speak it too loudly, and pursue vengeance with wrath-of-God vigor. Whatever the reason, the 1921 election for the presidency of the Geological Society of America was a case in point. At its April meeting that year, the society's governing council nominated its ticket of officers for the following year, including Schuchert as president. This unsolicited honor was "the greatest pleasure of my life," Schuchert later recalled, attained "on the basis of eminence achieved as a research worker." But it was also the occasion of his "greatest pain and disappointment."[11]

The society's custom then was for the official ticket to go unchallenged. But a petition soon emerged promoting an opposition ticket. The petitioners were ostensibly angered because the council's nominating process was undemocratic and because paleontologists were being favored over other specialties. But Schuchert's former mentor Edward O. Ulrich did all the preparation, writing, and mailing to make the opposition ticket happen, according to geologist Malcolm Weiss, who parsed the ensuing fight in a series of papers in the 1990s. Ulrich also carried the first copy of the petition by hand around the National Museum and other Washington, D.C., venues. "His was a powerful and domineering personality," Weiss wrote, "and a calm nerve was required to go against him." It took Ulrich just three hours to round up twenty-seven signatures. (While Schuchert immediately suspected Ulrich, who "is jealous of me and seeks my downfall," there were also others with whom "I have dared to differ scientifically.")[12]

Ulrich solicited additional support by mail, often sending a personal letter together with the official one, revealing the deep animosity that was the true motive for his campaign. To one correspondent, he wrote that Schuchert was "a man risen to undue prominence and arrogated leadership by insistence on receiving credit for ideas appropriated from others without mentioning their sources. As you know I have suffered much at his hands." He quickly added, "Please believe that I have entered this fight less on my own account than to support my friends who hold sore grievances against him." To another, he wrote that Schuchert's "favorite method" of getting what he wanted "is, and always has been, to use his friends and sponsors as stepping-stones and to leave them little but the scars of his heel-marks. I know because I raised and taught him and stood behind him until he committed the folly of thinking he no longer needed me."[13]

One correspondent wrote back, "Evidently you paleontologists are about as friendly as ever." Another suggested, reasonably, that the presidency had been due to go to an invertebrate paleontologist in any case. Having looked over the possibilities, the society's governing council had probably "found that the paleontologists were a scrappy bunch and that all had enemies." Schuchert merely seemed "the most available selection."[14]

Predictably, the opposition slate failed miserably, not least because the petitioners were unable to persuade their top two candidates to stand for office. In the end, Weiss concluded, the whole squabble came down to Ulrich's resentment over Schuchert's unwillingness to promote his grand scheme, his so-called Ozarkian and Canadian systems, possibly aggravated by Ulrich's having been obliged to borrow money from Schuchert over the years. Ulrich had acted out of the depth of his "hope to be immortalized" among the great names of his field "as the author of geologic systems." Instead, Ulrich's "Ozarkian and Canadian systems" have been entirely forgotten, except for "the petty jealousy and personal spite" of their author.[15]

Erased

For Schuchert, the other great sorrow of his career was the perennial threat to the Peabody Museum's continued existence. He soon became the heir to Marsh's endless efforts to give the "University a museum worthy of itself and of science," but without success over the course of twenty frustrating years. The move to the Hillhouse estate, which the Peabody trustees had accepted in 1905, was delayed for four years until the university found funds in 1909 to pay for the land, then by another six years due to lack of funds to build a new museum. The Peabody's building fund was still inadequate for the purpose on Christmas Eve, 1916, when Schuchert, then treasurer of the Peabody, received a visit from the university treasurer with tantalizing news: Anna M. Harkness, widow of one of the founders of Standard Oil, had donated a fabulous sum—$3 million—to memorialize her son, Charles W. Harkness, a Yale alumnus who had died suddenly that March. Schuchert's hopes must have soared as the treasurer outlined the university's proposed plan to triple the Peabody's existing $250,000 building fund to $750,000, with the total to accumulate at 4.5 percent interest until used. "It was truly to be a grand Christmas present for the Museum," he later recalled. Then, playing the part of wicked stepmother, Yale added one horrifying detail: construction of the Harkness Memorial Quadrangle, including the dormitories the college urgently needed, was to begin almost immediately on the land already occupied by the Peabody Museum. "To get this gift," the treasurer told Schuchert, "the Museum building must be erased and the property ready for the Memorial Quadrangle" by the following July, barely six months later.[16]

This badly flawed deal was formalized on March 19, 1917. Next day, Edward S. Dana began packing the minerals that had been the origin of the university collection and of the Peabody. "One after another all the staff followed his example," the curators wrote in their annual report, "and early in April all the collections were being packed so that they could be moved and stored in temporary quarters. The quiet old place of research and teaching was filled with the noise of packing, and a sadness came over the staff, for at least three of them had grown up with the collections." In Dana's case, this was literally so, and his father's benevolent ghost still seemed to preside as the ruling genius of the old building. "To see the splendid and rare specimens, large and small, dismantled and packed away in boxes and drawers," the curators continued, "took out of them for the time being most of their hopes of a greater museum."[17]

The university found room in the basements and attics of campus buildings and in a specially built storage space for twenty-five hundred boxes and upwards of twelve thousand drawers of material. The move began with plans already being drawn up for the new museum building. All found consolation in the thought that they would at least soon have space to display some of O. C. Marsh's huge dinosaurs and other Peabody treasures. But on April 6, 1917, just as this process was getting under way, the United States entered World War I, and costs for building materials and labor promptly soared. "It shortly became apparent that even the building of a new museum must be delayed," the curators reported.[18]

That August, Dana wrote advising Schuchert to stay away from New Haven and avoid the demolition, "which must make the Yale part of the town look a good deal like Belgium."[19] The staff was now scattered across the campus, and for the next nine years, as the university delayed and reneged on old commitments, fragile specimens deteriorated in overheated attics, damp basements, and temporary storage buildings. It was as if the museum made by George Peabody, O. C. Marsh, Benjamin Silliman Sr., James Dwight Dana, and the rest had never existed in the first place.

CHAPTER 17

A City Raised Like a Chalice

Shall we send out expeditions to explore unknown Peru,
When the cellar of Peabody offers work enough to do?
"The Solution," *Yale Alumni Weekly*, 1913

The Inca ruin called Machu Picchu—this "city raised like a chalice," as Chilean poet Pablo Neruda put it—is today one of the most popular travel destinations in the world. It has become "synonymous with Peru" and with the provincial capital of Cuzco, one anthropologist has written, "just as the pyramids have come to symbolize Egypt; the Parthenon, Greece; the Great Wall, China."[1] But Machu Picchu was little more than a rumor on July 24, 1911, the day an American explorer named Hiram Bingham III made this "lost city" visible to the world and to Peru itself.

Bingham, then thirty-five, was a Yale historian specializing in South America. He had started out with no particular interest in the Incas, focusing instead on the period of colonial domination. But in 1909, as a delegate to the Pan-American Scientific Congress, he traveled through Peru and was astounded by the massive drystone fortifications outside Cuzco. They were "the most impressive spectacle of man's handiwork that I have ever seen in America," he wrote. A few days later, a local official in the southern city of Abancay prevailed on a reluctant Bingham to break his itinerary and make the two-day climb up to a newly rediscovered Inca ruin called Choqquequirau, the "cradle of gold."[2] There was of course no gold there. But Bingham wasn't interested in gold. His imagination caught on the tantalizing possibility that there might be other such ruins out there still waiting to be discovered, and he began to make plans for his return.

Bingham was the son and grandson of missionaries in Hawaii. He had escaped

Hiram Bingham III worked hard to publicize his finds at Machu Picchu, inadvertently making the site the national symbol of Peru.

to Yale for college, to the University of California at Berkeley for a master's degree in history, and finally to Harvard for his doctorate, in the process slipping off the strictures of his hellfire-and-corporal-punishment childhood. He had married a Connecticut woman, Alfreda Mitchell, apparently infatuated as much by her plush life as heiress to the Tiffany fortune as by her more personal attributes. With the help of influential friends, he wangled an appointment as an assistant professor at Yale, unsalaried at first and never more than part-time. Then he settled into life with a growing family in a thirty-room mansion built by his in-laws on New Haven's Prospect Hill. Soon, though, he felt "the Bingham blood stirring in my veins," as he wrote to his father before his first expedition, to Venezuela and Colombia, in 1906, and was driven "to start out for little known regions as nearly all my Bingham ancestors for ten generations have done before me."[3]

Bingham fit the image of turn-of-the-century explorer almost to the point of parody. He was six feet four inches tall, with blond hair, blue eyes, prominent cheekbones, and a square jaw. In an expedition photograph, he stands beside his mule, peering out from under his battered campaign hat, unshaven, a neckerchief knotted at the throat, one hand on the horn of his saddle, the other cocked by one thumb in the hip pocket of his belted khaki explorer's jacket. Charlton Heston would recapture that image as Harry Steele in the 1954 film *Secret of the Incas.*

Harrison Ford would practically bottle and trademark it as Indiana Jones in the 1981 *Raiders of the Lost Ark* and many other films.

Bingham's 1911 expedition would be his first as organizer or, as he fashioned himself, "director" of a party of other researchers. An acquaintance from college, Edward S. Harkness, brother of the man whose death and resulting Memorial Quadrangle would soon bring about the demolition of the Peabody Museum, became an early supporter. To raise funds, Bingham also sold off the last of his family property in Hawaii and made a deal with *Harper's* magazine to write four stories about the expedition at $250 apiece.[4] Other alumni and corporate sponsors joined in, and the Yale Corporation agreed to put its name on what became the Yale Peruvian expedition of 1911. The party of seven ultimately included a surgeon, a geographer, a topographer, and a naturalist, all pursuing a complicated agenda that would send them off in multiple directions over the course of five months.

"I shall never leave you again," Bingham wrote to his wife, Alfreda, as he asked her to contribute the final $1,800 to his budget. "But I do believe that this next expedition will really add to Science and Truth. That will be your reward."[5] He seems not to have fathomed just how minimally rewarding this might seem to a wife left at home for half a year with a family then numbering six sons under the age of ten. (By 1914, it was seven.) Bingham would at least send many affectionate letters home. "Hiram was never more fond of Alfreda," their son Alfred later noted dryly, "than when he was away from her."[6]

Bingham's driving motivation wasn't simply the search for Incan cities, though that was the focus of Yale's interest. He hungered to be an explorer—it hardly mattered of what—and quoted Kipling to that effect:

> Something hidden. Go and find it.
> Go and look behind the Ranges—
> Something lost behind the Ranges.
> Lost and waiting for you. Go!

On his 1911 expedition, Bingham was less interested in Incan ruins than in becoming the first person to climb the volcanic peak Coropuna, according to his son Alfred in his 1989 biography *Portrait of an Explorer.* The American mountain climber Annie S. Peck had just made the first ascent of Peru's Mount Nevado Huascarán and published a book, *A Search for the Apex of America.* Bingham believed Coropuna, to the south, might actually be higher and, according to his son, he wanted "to assert the superiority of the male in the face of the presumptions of feminism." He would have his chance: on the last leg of the Bingham party's voyage south along the Pacific coast to Peru, Peck turned out to be a fellow passenger also en route to Coropuna. Bingham found her to be "a hard faced, sharp tongued old maid of the typical New England school marm type. She must be at least 55."[7] In fact, she was sixty, and the more intrepid for it.

For Bingham, though, Coropuna was still months away, the last item on his itinerary, even if it was the first item for Peck. He was headed now to Cuzco and then into the Urubamba Valley. His Peruvian informants had told him that this was where he was most likely to find remnants of Incan culture, possibly including the legendary last holdout of the Incan emperors, called Vitcos or Vilcabamba. A German travel book, "picturesque but unreliable," according to Bingham, had noted the rumor of such ruins as far back as 1875.[8] By 1911, though, access to the valley had become much easier because of a new government road for mule trains on the riverbank. One day, in the course of organizing logistics and packing expedition members off on their various assignments, Bingham took that road out to the small town of Urubamba. There a local official provided him with his most definitive clue: at a mountain called Huayna Picchu, a short distance downstream, he would find ruins better than the ones he had seen in 1909 at Choqquequirau.

Soon after, on his way into the valley, Bingham encountered the expedition's topographer, who had gone ahead. He had grim news: an Indian boy hired to carry surveying equipment across the river had fallen in the rapids and drowned. Bingham made no mention of the incident in his journal. "Perhaps he felt Indians were more expendable than surveying instruments," his son wrote, "and the instruments, though damaged, had been saved."[9] Instead, he pressed on to his destination, below the nine-thousand-foot peak of Huayna Picchu. (The name meant "new peak," as opposed to its neighbor Machu Picchu, "old peak.")

A local muleteer named Melchor Arteaga agreed to serve as a guide, and he and Bingham began their climb at 10:07 a.m. on July 24. At 10:40, they paused long enough to take a photo with Machu Picchu visible at the top. Bingham noted a dead snake by the roadside. It would become the basis for transforming the area, in standard adventure-writing style, to "the favorite haunt of 'vipers'"—specifically, venomous fer-de-lance snakes, "capable of making considerable springs when in pursuit of prey." Arteaga and Bingham came to a bridge made of long, thin tree trunks lashed together with vines. At the middle, it bent beneath them to within a few feet of the roaring Urubamba River. From there, they made a "fearfully hard" climb, sometimes scrambling upward on hands and knees, clinging to roots and tree trunks for traction in the drizzling rain.[10]

The climb took just two hours, and they were welcomed to the saddle of land at the top by three farmers who had recently moved there to avoid government agents seeking military recruits, among other intrusions. But they were friendly

The overgrown ruins of Machu Picchu at the time of their 1911 rediscovery. Anthropologists now think the site was a country retreat, the Camp David of Incan emperors.

enough, and the two visitors gratefully accepted the gourds of "cool, delicious water," and the boiled sweet potatoes they offered. Bingham then continued farther on for a few minutes, a farm boy showing him the way, while Arteaga rested. The pencil-scratched handwriting in the little leather-bound notebook Bingham carried became a bold scrawl: *"Fine Remains—much better than Choq."*[11]

The farmers had burned away enough of the tangled overgrowth to make visible what Bingham described in a 1913 article for *Harper's* as a "maze of small and large walls, the ruins of buildings made of blocks of white granite, most carefully cut and beautifully fitted together without cement." They were arranged around an open plaza on a ridge connecting Huayna Picchu and Machu Picchu, with neat agricultural terraces marching up and down the slopes at either end. Bingham called them "as wonderful ruins as any ever found in Peru."[12] He seems, however, not to have recognized in 1911 just how wonderful they were. He took a few pictures on that first visit, prowled around the buildings, then turned around and headed back downhill, arriving in camp before dark, like any modern tourist.

The rest of his itinerary beckoned: just two weeks later he would locate the remains of Vitcos, the last holdout of the Incan emperors, an achievement almost on a par with finding Machu Picchu (though he did not recognize the significance of this discovery either). The expedition also needed to complete a survey of the 73rd meridian, roughly from Cuzco down to the Pacific Ocean, and of course Bingham meant to climb Coropuna. (Peck got there first, although Bingham seems to have climbed higher. It is doubtful that either of them reached the summit, which was not, in any case, the apex of the Americas.) The result was that Bingham himself did not bother to return to Machu Picchu that year, though he sent a two-man crew up to map what he described to his wife as "my new Inca City." He thought the expedition's greatest discovery was a site near Cuzco where he had found intermingled human and animal bones embedded in the exposed face of a hillside beneath seventy-five feet of gravel. Bingham theorized that his "Cuzco man" might indicate a much earlier human presence in the Americas than anyone had ever imagined.[13]

Burying Cuzco Man

Bingham would return to the Urubamba Valley in 1912 and again in 1914–15 on expeditions jointly sponsored by Yale and the National Geographic Society. The 1912 expedition for the first time included a scientist from the Peabody Museum, assigned to determine the antiquity of the "Cuzco man" site. Bingham and osteologist George F. Eaton had previously identified one bone from the site as belonging to a species of bison not previously known in South America. But on the scene, Eaton quickly realized that it came from a kind of cattle still being sold in the Cuzco market. He also recognized that the mixed bones found beneath the gravel were relatively modern graves, burials in cells that had been scooped out of the hill face, the dead provisioned with the meat of various animals to sustain them in the afterworld. On hearing Eaton's conclusions, a crestfallen Bingham fired off telegrams to prevent further publication of his now embarrassing "Cuzco man" speculations. According to Alfred Bingham, he never forgave Eaton for debunking what he had imagined to be his greatest discovery.[14]

Instead, Bingham eventually shifted his emphasis to the discovery of Machu Picchu. He settled on the theory that it was the alpha and the omega of Incan civilization, both its mythical birthplace and also its final refuge after the Conquest. (He still discounted his find at Vitcos, the actual last refuge.) Eaton supervised the excavation of graves at Machu Picchu, concluding that the population had been predominately female, a group he and Bingham came to think of as—what else?—"Virgins of the Sun."[15]

Eaton mapped and made detailed notes about whatever he excavated, and in December 1912, the expedition shipped about one hundred crates of specimens back to the Peabody Museum. The contents were mostly bone and pottery frag-

are hard to come by. Richard Burger and Lucy Salazar, whose mixed backgrounds and marriage would have made them unwelcome on any Bingham expedition, found themselves rising to Bingham's defense.

They argued that Bingham's critics were suffering from acute "presentism," the tendency to distort the past by anachronistically applying present-day values and perspectives. To complain that Bingham lacked archaeological training, for instance, ignored the historical reality that most "major archaeological figures of the early 20th century had been trained in other disciplines" and that "professional archaeologists were exceedingly rare" then. Bingham's 1912 archaeological research was in many ways "far ahead of its time," they wrote, notably for "his meticulous mapping [of] the site using state of the art cartographic equipment" and for "his sampling of the different sectors of the site in order to infer differences in function and activities carried" on there.[25] The value of his collection did indeed reside in quantity: the common archaeological practice at the time would have been to discard random animal bones and metallurgical residues. Bingham preserved them all, greatly increasingly the context of the collection for later researchers. He also pushed out beyond the ruins themselves to locate the old Incan road system linking Machu Picchu with nearby communities.

The implication that Bingham and the Peabody Museum had somehow been on a mission to fill their coffers with the treasures of Peru also missed one painfully obvious fact: at the time of Bingham's expeditions, the Peabody was expecting at any moment to be demolished. It already had collections stacked up in their unopened crates because it lacked the space to store them. The only riches it needed then were the kind that a donor might have provided (but so far hadn't) to build it a new home. The Machu Picchu artifacts as physical objects were never really the point, for Bingham or the Peabody, according to Burger and Salazar. Bingham was more excited by his photographs and by the stories he could tell a popular audience. The Peabody was engaged by the potential to produce science, in books and in scholarly journals. "This was true at Yale in 1912," Burger and Salazar wrote, "and remains true today."[26] Recognizing this emphasis on scholarly research over physical possession would ultimately be the key to the settlement of the Machu Picchu fight.

Late in 2010, Yale and Peru announced a joint agreement. As "an expression of good will" and in recognition of "the unique importance that Machu Picchu has come to play in the identity of the modern Peruvian nation," Yale would return the entire Machu Picchu collection.[27] Peru in turn acknowledged Yale's role in the care and preservation of the collection. The University of Cuzco (Universidad Nacional de San Antonio Abad del Cusco, or UNSAAC) would create a museum space where the collection would remain accessible to scholars. It would also provide for the conservation and security of the collection and display suitable items to the broadest possible public. The University of Cuzco and Yale would administer the museum together through the U.N.S.A.A.C.–Yale International Center for the Study of Machu Picchu and Inca Culture.

The first shipment back to Peru arrived in June 2011, in time to mark the hundredth anniversary of Bingham's find. A crowd of fifteen thousand people lined the streets to welcome home objects that had seemed worthless on their departure from Peru a century earlier but were returning now almost as relics.

The museum soon opened in downtown Cuzco in a handsome colonial-era building. It was an architectural as well as an intellectual joining of two worlds: Casa Concha stands on the foundations of the palace of the great Inca emperor Pachacuti, and the first thing visitors see on entering is a diorama of the 1912 excavation of Pachacuti's greatest monument. It features a life-size figure, in campaign hat and belted explorer's jacket, of Hiram Bingham III, now honored in Peru as the "scientific discoverer" of Machu Picchu.

CHAPTER 18

Teaching Evolution

[The creationists] cannot hurt evolution, but will do much harm in their following in bringing on arrested mental development.

CHARLES SCHUCHERT to George MacCurdy, 1922

On December 29, 1925, more than eight hundred people crowded into the Great Hall of Dinosaurs for the dedication of the Peabody Museum's new building. Paleontologists, geologists, mineralogists, anthropologists, and zoologists, who had all scheduled the yearly meetings of their professional societies to coincide with the event, packed in close around the *Stegosaurus* skeleton at the center of the hall. Wieland's giant *Archelon* sea turtle skeleton rose up out of a sea of balding heads and fedoras. *Claosaurus* (*Edmontosaurus*) ran behind the speaker's lectern, eternally trapped by sandstone in midstride. Marsh's *Brontosaurus,* for which this hangarlike space had been designed, would need a few more years of preparation before it could rise up here and stretch out to its full seventy-foot length.

It was a good thing that Marsh himself could not be present, because the keynote speaker was his old nemesis Henry Fairfield Osborn, now in Marsh's former place as North America's leading paleontologist. "The opening of this superb museum marks an epoch in the development of Vertebrate Paleontology in America," Osborn declared. He had enough sense of occasion to add that it was "inspired by the monumental labors of Othniel Charles Marsh" and "filled with Marsh's unique collections." He acknowledged that Marsh had made the Peabody "the mecca for the evolutionists of Europe" and Yale University "the most famous center of vertebrate paleontology in the world." Osborn could not resist using his term *titanotheres* rather than Marsh's *brontotheres,* and he went on to reunite Leidy, Marsh, and Cope as the great trinity of nineteenth-century American paleontology. The

At the opening of the new Peabody Museum building in 1925, the audience crowded in around the skeletons in the Great Hall. (The *Brontosaurus* for which this space was designed was still in the planning stages.)

heart of his talk, though, was a defense of evolution, including a proposal that scientists stop talking about evolutionary theory and talk instead about the *law* of evolution, like the law of gravity.[1]

The *New York Times* devoted almost a quarter page in its news section to the event, emphasizing the beliefs of those present: "A large proportion of them, perhaps a majority, are church members or active in church work, and all of them are believers in evolution. Close inquiry today did not reveal an atheist among the several hundred scientific men." A subhead drove home the point: "Scientists All Religious."[2]

With the same idea of breaking down any apparent divide between theology and evolution (or between institutions), Charles Schuchert quoted a poem originally written for the Academy of Natural Sciences of Philadelphia:

Great God of nature, let these halls
The hidden things of earth make plain;
Let knowledge trumpet forth her calls,
And wisdom speak, but not in vain.[3]

To himself, he must have muttered a simpler prayer on finally seeing the scattered Peabody collections united again under one roof: "*Thank God.*"

Back from the Wilderness

Getting to this moment had been a struggle. As head of the Peabody during much of its exile, Schuchert had pleaded and pushed for years to get the university to fulfill the commitments it had made in rushing to demolish the old museum. But the Peabody had somehow become "the goat" and "the step child of the University." Yale had promised to add $500,000 to the $250,000 accumulated in the Peabody building fund, the total growing at 4.5 percent. But "all the money we have is what we had in 1917," Schuchert wrote in 1922, "and President Angell will not allow us to build up to the limit of our funds. In consequence we are not going to have the space hoped for."[4] Angell wanted to withhold a large chunk of the building fund as an endowment to pay for the museum's operating costs, though Schuchert repeatedly protested that this violated the 1917 contract. Two members of the Yale Corporation somehow even figured the Peabody construction should be whittled down to just $211,000.

Nine years "of gloom and confinement," as Schuchert termed it, had ultimately cost the Peabody its research and educational momentum, moved it more than a mile from the center of the campus, and in return gained it less than an acre of added room. Schuchert had told the university that the museum required 70,180 square feet of space at a bare minimum. Instead, it got just 58,000, up from 34,000 in the old building, and not all of that would be useful. The Yale Corporation had rejected the museum's original architectural plan because it lacked a "strategic design" for its corner location, meaning it needed a tower. "Beautiful things, however, cost lots of money," wrote Schuchert, who figured this would eat up another $150,000 from the building fund, with only a third of the tower space "of value to the museum curators." The university's architectural planning committee had also simply deleted without explanation the 3,600 square feet for display of anthropology collections: "I know that you will be disappointed," Schuchert wrote anthropology curator George G. MacCurdy in 1922, "but in the great squeeze, because of lack of room, all are disappointed and you the most."[5]

Schuchert found the continued neglect more galling because New York attorney John W. Sterling had left his entire estate to Yale on his death in 1918, a bequest that would eventually total $29 million. The Peabody had originally been in second place for these funds, after completion of the university's new Sterling Memorial Library, Schuchert wrote, only to see new buildings for chemistry and the medical school take its place, leaving the Peabody "outside the Sterling money!" He added, "Our being pushed aside for other departments has left us in great discouragement."[6] Fed up, he resigned all administrative duties in 1922,

The current museum building under construction in 1924. Peabody staff proposed a major expansion just five years later. But by 1930, the Depression had begun.

nominating Richard S. Lull, who had been a vertebrate paleontologist at the Peabody since 1906, to take his place.

It's hard to say why the university was so stingy and slow, even mean-spirited, with the Peabody. Maybe natural history seemed out of sync with the growing emphasis on genetics, experimental biology, and other laboratory sciences. The excitement about dinosaurs may also have died down as they came to be deemed an evolutionary dead end, a one-word symbol of everything old and unworkable. The museum may also have been an easy target for the perennial hostility to science within the humanities faculty. What the Peabody needed—but lacked—was, as always, a Benjamin Silliman Sr., or even a Henry Fairfield Osborn, to make its case, to woo wealthy donors, and to remind Yale that the museum's mission was at the heart of the university—and of the Earth.

In the face of these challenges, the Peabody could easily have retreated. Instead, it took an extraordinarily bold step, remaking itself as a powerful American voice on behalf of evolutionary thinking—even planning to teach it to schoolchildren—at the very moment that Darwinian teaching was being criminalized by fundamentalists clinging to their Bibles. The Peabody made this stand in the most forthright possible terms. The headline of the *New York Times* report on the opening ceremony made that clear. This was no longer just about the arguably quaint business of natural history. Instead, the *Times* announced: "Evolution Museum Dedicated at Yale."

The Monkey Trial

Seven months earlier, on May 5, 1925, the state of Tennessee had charged a substitute science teacher with contradicting the biblical account of human origin, in violation of a new state law. William Jennings Bryan rushed to take up the case against "this slimy thing, evolution."[7] He was a three-time Democratic nominee for president, an anti-bank populist, and a peace advocate who lobbied on religious grounds for state laws, and even a constitutional amendment, banning the teaching of evolution in schools. Clarence Darrow, the leading defense lawyer of the day, agreed to represent both science and twenty-four-year-old John T. Scopes, who had become the defendant after volunteering to serve as a test case for the new law. What became widely known as "the Monkey Trial" was about to put evolution and the small town of Dayton, Tennessee, on the front page of every newspaper in the country.

With the trial due to begin, *Time* magazine featured evolution on its cover, in the person of paleontologist and Peabody Museum director Richard S. Lull. In the cover artist's sketch, Lull stared out placidly through pince-nez glasses, a calm, even reassuring figure, with a high balding forehead, a neat mustache, and a narrow white flame of beard extending down from his lower lip. Below his name was the phrase "of a very ancient lineage." This was a reference to Lull's assertion that humanity was "not the result of instantaneous creation" but "comes of a very ancient lineage," the product "of an orderly and long-drawn-out evolution." The writer went on to depict Lull as a sort of evolutionary paragon, "over 6 ft. in height, sturdy, straight as an arrow, with regular features, a high arching forehead, a keen mind, soft spoken (although suffering, like Edison, from deafness), courteous, kindly, possessed of a sense of humor—all the attributes commonly thought of as the height of human attainment."[8]

This was of course a lot for any mortal to live up to, and Lull could also be petty and self-important. Once, at a Peabody staff event, the person introducing a talk by Schuchert identified him as the museum's former director. Lull immediately interrupted to point out that Schuchert had never held that title, which did not exist until Lull's appointment in 1922. The long-suffering Schuchert stood by, not bothering to clarify that the title only existed because he had lobbied over considerable opposition to create it for Lull.[9]

Even so, Lull was a highly successful teacher at Yale, and his course on organic evolution was "one of the most popular ever given at that university," his former student George Gaylord Simpson later wrote. Hundreds of students each year

As schoolteacher John Scopes was facing trial for teaching Darwin, *Time* magazine made Peabody Museum director Richard Swann Lull the face of evolutionary thinking.

were drawn in by his "impressive bearing, his skilled delivery, and his complete command of his subject." Lull's *Organic Evolution* was the standard college-level text and, with impeccable timing, his book explaining evolutionary thinking for the general reader arrived in bookstores a month before the Scopes trial was to begin. The *New York Times* praised Lull's *The Ways of Life* for "meticulously distinguishing between the proved and the hypothetical, perhaps the unknowable, never claiming for his science more than extensive, tested data justify."[10]

Lull now looked like the perfect expert witness, both for the judicious tone of his book and for his ringing conclusion: "Direct creation is but a bit of ancient folklore" with "roots in the stories told by long-forgotten sages to dimly remembered peoples." On July 10, as the Scopes trial was getting under way in Tennessee, the defense sent him a telegram:

WESTERN UNION TELEGRAM

NEWCOMB CARLTON, PRESIDENT GEORGE W. E. ATKINS, FIRST VICE-PRESIDENT

Received at 85 Orange St., New Haven, Conn. ALWAYS OPEN

1925 JUL 10 AM 4 34

NA38 72 NL

DAYTON TENN 9

DR RICHARD S LULL

DEPT OF PALEONTOLOGY YALE UNIV NEWHAVEN CONN

DISTINGUISHED COLLEAGUES OF YOURS HAVE SUGGESTED YOU MIGHT BE WILLING TO COME TO TESTIFY FOR DEFENSE AT DAYTON TENNESSEE NEXT WEEK IN THE CASE OF STATE OF TENNESSEE VERSUS PROFESSOR SCOPES STOP WE OF THE DEFENSE WOULD BE DELIGHTED TO ADD YOU AUTHORITY TO OUR POSITION STOP YOUR EXPENSES WILL BE PAID STOP WILL YOU WIRE ME DIRECTLY AT DAYTON AND I WILL LET YOU KNOW WHAT DAY YOU WILL BE NEEDED

CLARENCE DARROW.

By telegram, defense attorney Clarence Darrow asked Lull to testify at the Scopes evolution trial in Tennessee.

> DISTINGUISHED COLLEAGUES OF YOURS HAVE SUGGESTED YOU MIGHT BE WILLING TO COME TO TESTIFY FOR DEFENSE AT DAYTON TENNESSEE NEXT WEEK IN THE CASE OF STATE OF TENNESSEE VERSUS PROFESSOR SCOPES STOP. . . . WIRE ME DIRECTLY AT DAYTON AND I WILL LET YOU KNOW WHAT DAY YOU WILL BE NEEDED
> CLARENCE DARROW.[11]

But Lull was away teaching two summer courses at the University of California at Berkeley. He would in any case have been horrified by the circuslike spectacle of the trial, with banners near the courthouse advising, "Read Your Bible," "Prepare to Meet Thy God," and "Repent or Be Damned."[12]

The trial lasted just over a week. The most dramatic moment came when Darrow, the "infidel," as he was known, put Bryan on the stand and questioned him point by point about his belief in various implausible notions in the Bible. Worn down and exasperated, Bryan finally burst out, "I do not think about things I don't think about." Darrow replied, "Do you think about the things you do think about?" With spectators laughing at his humiliation, Bryan answered haplessly, "Well, sometimes."[13]

The newspaper columnist and pundit H. L. Mencken ridiculed Bryan as "this old buzzard" and a "buffoon." But Bryan was also a canny lawyer, and he was soon arguing to the judge that no scientific evidence was needed to decide the case. The law was clear, and all that mattered was whether John Scopes had taught evolution, yes or no. The judge responded, in Mencken's words, by "leaping with soft judicial hosannas into the arms of the prosecution." He struck down the testimony of the one scientific expert who had already appeared and refused to allow any others to take the stand. Scopes was fined $100, and the law against teaching evolution remained on the books until 1967. Bryan, victorious, died five days later. Mencken had the last word: "God aimed at Darrow, missed, and hit Bryan instead."[14]

The most enduring result of the trial was its devastating effect on the teaching of evolution, detailed in a 2001 article by University of Minnesota biologist Randy Moore. In the immediate aftermath of the trial, the governor of Texas ordered discussion of evolution to be cut out of high school textbooks with scissors. Timid publishers, who had previously featured evolution as the foundation of modern biology, took this as a message aimed directly at their pocketbooks. One textbook that had previously carried a portrait of Darwin on its frontispiece appeared in its 1926 edition with a drawing of the digestive tract in his place. Bryan had complained before the Scopes trial that he could not find "any text book on biology which does not begin with monkeys." But by 1929, a fundamentalist preacher could happily declare that "virtually all textbooks on the market have been revised to meet the needs of the Fundamentalists."[15]

Telling a Different Story

This was the moment in which the Peabody Museum chose to open the doors of its new home with its displays now arranged, the *New York Times* reported, to show "the progress of animals from mud to man"—that is, with human origins and progress "set forth merely as a part of the general evolution of the animal kingdom." Just to be perfectly clear, the *Times* noted, "In the portrayal of man's origin the Garden of Eden will be absent." Lull was the driving force behind the decision to create a "museum of ideas," as an early Peabody guidebook put it. But the invertebrate paleontologist Carl Dunbar also made extensive scouting forays to other natural history museums for hints about how to present those ideas—and how not to: "The specimens," he wrote in his notebook after one such visit, "are simply put out like lines of canned goods on grocery shelves with the scientific names & geologic horizon. Nothing to teach a lesson to the public!" Of another, where he deemed the displays "no more interesting than a dictionary," he added emphatically, "About 9/10 of the material exhibited is of little interest *because it doesn't tell a story.*"[16]

The Peabody was determined to tell a story—a dramatically different alternative to the story of Genesis. The museum's first-floor layout was designed to lead visitors in a logical sequence through five major halls. First was a large room dedicated to invertebrates — among them Beecher's trilobites, Schuchert's brachiopods, and Verrill's giant squid. Then came fishes and reptiles and a display of *Ichthyornis* and bird evolution, all arranged around the perimeter of the Great Hall of Dinosaurs. The dutiful visitor was then meant to visit the first Hall of Mammals (even-toed hoofed mammals, or *Perissodactyla,* with an emphasis on the evolution of the horse) and the Second Hall of Mammals (*Artiodactyla,* from the Irish elk to the marine mammals, together with some carnivores and a few early primates). The climax of this story was the Hall of Man where, as an early guidebook described it, "the steady upward progress of life" achieved its grand result in *Homo sapiens.*[17] This emphasis on progress reflected Lull's belief in the "orderly" nature of evolution, with a self-improving neo-Lamarckian strain added in to mellow the blindly random mutations of Darwinian natural selection. But there was no soft-pedaling the big conclusion: in one dramatic display, a sequence of five skeletons led, as if step by step, from gibbon to orangutan to chimpanzee to gorilla to man.

The *New York Telegram-Mail* was horrified. It called on antievolutionists to take note of "the insidious propaganda going on in the Peabody Museum at Yale." Worst of all, the editorialists fulminated, "man in his majesty is not to be exalted above the beast." It wasn't clear "just what can be done" about "the plotting of the Yale scientists." But in other states, the editorial hinted, "Mr. Bryan has sponsored legislative action to prevent such teaching." Syndicated columnist Arthur Brisbane was also outraged: "The Peabody museum at Yale — please shudder — contains an exhibition purporting to prove the evolution theory, showing our 'biological ascent from protozoa to man.' To be told that you are descended from monkeys is bad enough, but ten thousand times worse is the idea that your ancestors include the protozoa. Anti-evolutionists would be ashamed to recognize such relatives, of whom any number could live happily in a thimbleful of salt water."[18] Altogether, the opening and the evolution-driven displays generated almost three hundred newspaper headlines, from San Francisco to Greenville, South Carolina.

Explicitly embracing the evolutionary perspective was undoubtedly a bold step in that era. (To put it in perspective, at least one major North American natural history museum still finds the issue problematic in the twenty-first century, avoiding any overt mention of evolution, a spokesperson explained in 2014, because "We don't need people to come in here and reject us.") But the conspirators of the Peabody Museum had an even more nefarious objective in mind. As the *Springfield (Mass.) Republican* reported, in a swoon, "Evolution is to be taught to groups of children who visit the new Yale Peabody museum." Yes, schoolchildren. The

Thinking with the Seat of Its Pants?

For a brief, shining moment, until the fully mounted *Brontosaurus* shoved it aside in 1931, *Stegosaurus* held center stage in the Peabody's new Great Hall of Dinosaurs. This bizarre seven-thousand-pound dinosaur has also played a curious role—or so it seemed until recently—in a beloved bit of light verse to which the Peabody Museum was an important offstage contributor.

The story starts with an article O. C. Marsh published in 1881 describing a "large vaulted chamber" in the vertebrae of the sacrum at the base of *Stegosaurus*'s spine. Marsh had already remarked on the incredibly small three-ounce brain—and apparent stupidity—of this massive dinosaur. Now, finding a neural cavity at the tail that was "at least ten times the size of the cavity which contained the brain," Marsh hit on the idea that it was a "posterior brain-case." Or, as the *Chicago Tribune* newspaper

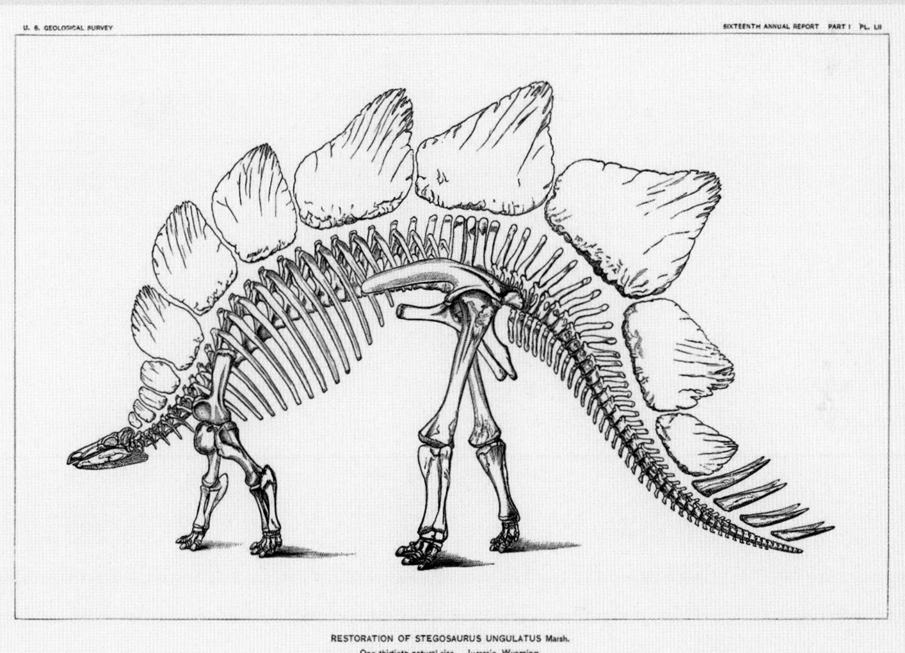

Marsh briefly mistook the jumble of bony plate fragments from an excavation for evidence of a giant marine turtle. But better and more complete specimens enabled him to create this iconic reconstruction of *Stegosaurus*.

columnist Bert Leston Taylor put it, in the best-known lines of his 1903 poem "The Dinosaur":

> The creature had two sets of brains—
> One in his head (the usual place),
> The other at his spinal base.
> Thus he could reason *a priori*
> As well as *a posteriori.*

The only hitch, as geology writer David B. Williams revealed in a 2013 article, is that the poem wasn't actually about *Stegosaurus.* Taylor was inspired instead by an article that had appeared in the *Tribune* the previous day about a seventy-foot-long *Brachiosaurus* specimen, newly discovered by Elmer S. Riggs of the city's Field Museum of Natural History.[a]

So how does the Peabody fit into the story? The poet made his source clear in a line beneath his title: "According to Prof. Farrington of the U.C. [University of Chicago]." And the previous day's *Brachiosaurus* article included a photo of Oliver C. Farrington poised with hammer and chisel to remove substrate from a massive and conspicuously hollow vertebra. The caption revealed that this dinosaur also had a second brain "sixty feet from the primary seat of his intelligence." Farrington had apparently picked up Marsh's "posterior brain-case" idea as a doctoral student in geology at Yale—and then simply applied the idea to *Brachiosaurus,* too.[b]

Scientists now know—*a posteriori,* so to speak—that neither dinosaur had a brain in its backside. Instead, the vertebral cavity probably served as an energy reserve for the nervous system, like the glycogen body in the spinal cords of modern birds. Contrary to Taylor's playful notion that such dinosaurs "could think, without congestion, / Upon both sides of every question," the evidence suggests, sadly, that they barely needed to think at all.[c]

The layout of the museum led visitors on an evolutionary tour culminating with a frank emphasis on the close connection between other primates and human beings.

New York Telegram-Mail added, "Thousands upon thousands of innocent young school children are to be inveigled into the museum, there to have their young and impressionable minds steeped with evolutionary doctrines."[19]

The Peabody had created a school service division, among the earliest of its kind; this was almost as radical a departure from the conventional practice of university museums as its storytelling. The decision to carry its case to young people came from Lull, together with Yale president James R. Angell, who wrote to him of "very definite and great educational obligations to the interests of the general public and particularly to the secondary schools." That unwanted space in the museum's new tower would accommodate a classroom where trained staff would "make available the rich stores" of the museum "to the public school children of the State." (Women with advanced degrees were, for once, to be in charge.) Connecticut was of course not Tennessee or Texas, and its public officials welcomed this move. The state commissioner of education became a trustee of the Peabody Museum and presented a typed page of ways he thought the Peabody could be

Evolution and Other People's Morals

A common objection to teaching evolution, in the early twentieth century as now, was that it would somehow lead to immorality: in effect, telling people about their origins in the animal world would cause them to behave like animals. But Yale was, if anything, excessively concerned with morality then, to its own detriment.

When George Gaylord Simpson was a senior at Yale, he violated the university's policy against undergraduates being married, secretly wedding a Barnard College student named Lydia Pedroja. Simpson went on to earn his Yale doctorate in paleontology under Richard S. Lull. In 1927, while he was in London for a further year of study, he became a candidate for a position at Yale. But then Lull showed up, inquiring into a malicious report that Simpson was living in luxury while failing to provide for his wife. Simpson explained that Pedroja had refused to live in London, taking their two daughters to the south of France, and that he was supporting them with considerable difficulty. His "luxury" living, meantime, was in a garret. Yale soon confirmed its job offer, but Simpson "declined to go where prospective associates had been so willing to believe me a scoundrel."[a]

In the mid-1930s, Simpson and Pedroja separated and, pending their divorce, he began to live with his childhood friend and future wife, Anne Roe. By then well established in a career at the American Museum of Natural History, Simpson necessarily kept his living arrangements hidden. But Yale still worried about his moral character. Thus Simpson was put in the position of lying "when asked directly by Carl Dunbar, a Yale colleague, if he was living with Anne," according to Simpson biographer Léo F. Laporte. When "Simpson said, 'definitely not,' Dunbar replied that he was sure that Simpson 'would certainly do nothing of the sort.' This falsehood still bothered Simpson in his old age," according to Laporte, who interviewed Simpson in 1982.[b]

Simpson continued to flirt with the idea of joining the faculty at Yale, and Yale later presented him with an honorary doctorate as well as the Peabody Museum's Addison Emery Verrill Medal. But he spent his career elsewhere instead. Yale thus lost out on one of the leading figures in twentieth-century paleontology and a pioneer of the modern evolutionary synthesis. For Yale evolutionists, it was a self-inflicted wound, caused at least in part by too much concern, rather than too little, for what passed at that time for morality.

Lull and Gibb put the resurrected *Brontosaurus* in perspective. An ambitious mural would later re-create the dinosaurian world on the blank yellow brick wall in the background.

useful to schools. Among his suggestions: "Museum studies of mother care—birds, animals, man," "Museum beginnings I might make at home," and "Museum evidences of evolution: plant, animals, birds—etc. principles." (That "etc." must have been his way of saying "humans, too.") The Peabody's school service program also taught a conservationist ethos, another unusual initiative for museums then. "Some exhibits are intended to make the children familiar with the wild life of this region and to encourage them to protect it," the *New York Times* reported. "Others show the harmful results produced by illegal methods of obtaining birds or their plumage for commercial purposes." Finally, "Particular use will be made of the rare material on American Indians which composes the major portion of Yale's ethnological collection."[20]

Local schools responded eagerly. In the first month alone, twenty-nine classes from public and private schools came to the museum, and demand increased to the point that, by mid-1927, the Peabody's executive committee reported, "Further expansion of work of department hardly possible under present conditions." The following year, the school service program moved into a separate building around the corner on Hillhouse Avenue. The rest of the museum was equally crowded. That same report, written at the end of the new building's first full year of operation, cryptically noted, "Realisation present space must suffice" and "*Storage*—nearly saturated, so that extensive additions to collections cannot be cared for without further space." But 74,467 visitors had come through the doors and, for the first time in many years, "Esprit de corps excellent."[21]

No one seems to have flung paint on an offensive display case or shouted biblical passages in the hallways. But the Peabody's evolutionary scheme would ultimately face a larger and more intractable challenge than anything the disciples of William Jennings Bryan could have dreamed up. Lull had thought the museum's new plan was so carefully laid out that "a casual visitor is automatically routed through the exhibits in a natural order," on a circuit from "the beginnings of the ascent of life" in the hall directly opposite the front doors and continuing all the way around the exhibit rooms to "its climax in the Hall of Man," returning through a door on the right side of the lobby.[22] But in 1936, a doctoral candidate named Mildred C. B. Porter made a study of how Sunday-afternoon visitors actually moved through the museum and discovered that people are less susceptible than imagined to automatic routing.

The planners had understood evolution well enough, but they were evidently less familiar with human nature. Porter summarized the result in a single devastating sentence: "When the individual records were examined, it was found that the route taken by the average visitor was the reverse of that planned in the Guide Book, 24.4 percent of the exhibits were examined, 10.9 percent of the labels were read, and the average time taken by the visitors for reviewing the history of life on the earth during the past 500,000,000 years was 21.40 minutes."[23]

CHAPTER 19

The Rise of Modern Ecology

I had never realized what an extraordinary number of pigeons are bright green.

G. EVELYN HUTCHINSON, *The Enchanted Voyage*

In 1928, a young Cambridge University graduate with no doctoral degree, who had failed in his first real experimental research at the Stazione Zoologica in Naples and been fired from his first job as a lecturer at South Africa's University of the Witwatersrand, arrived sight unseen for a temporary teaching position at Yale. There was little to indicate that he would remain beyond a brief fellowship, much less for the next sixty-two years (counting his continued work in retirement). Nor was there any reason to suspect he would become, in biologist E. O. Wilson's words, the "founder" of modern ecology.[1]

G. Evelyn Hutchinson had effectively chosen his profession by the age of five, when he was already collecting sticklebacks, water spiders, and water mites to keep in small aquariums. His parents—a mother who was "a strong but strictly nonmilitant feminist" and a father who taught mineralogy at Cambridge—indulged his interest in all things aquatic. Up to a point: they presumably drew the line when he pushed a friend into a pond at the University Botanic Garden "to see if she would float, an early interest in the hydromechanics of organisms of which I am not proud."[2] In the same spirit of experimentation, he and his younger brother once shut off the electricity and locked their parents—and their parents' guests, Sir George and Lady Darwin—in the dining room during a dinner party. (No information on the outcome of this experiment survives.) Hutchinson went on to a preparatory school with a reputation for science and there, at age fifteen, he published his first paper, about a swimming grasshopper, followed by another at sixteen, about a rare species of water strider (a type of insect, also called Jesus bug,

G. Evelyn Hutchinson included these European magpie moths in an influential essay about nineteenth-century contributions to genetics by amateur lepidopterists.

that skims across the water surface). Though he intended to study zoology, he took his entrance exams at Cambridge in mathematics, physics, and chemistry. He would combine zoology with these disciplines, hitherto largely foreign to natural history, for the rest of his career. It was the key to all his future discoveries.

A young Hutchinson takes to the field with his butterfly net in the 1920s.

The failure in his research at Naples was largely a result of being unable to obtain enough specimens of his study animal, an octopus, and perhaps also of the misguided decision to undertake a marine project: he was by nature a freshwater biologist. He was likewise unsuited to his task during a two-year contract as a lecturer at the University of the Witwatersrand. The professor to whom he was assigned there was accustomed to having his lecturers function in the medieval manner as readers of lectures he had written. Hutchinson apparently viewed the job differently, with the result that he was soon dismissed from teaching. He spent the remaining time on his contract studying South African birds and other wildlife, and especially increasing his knowledge of water bugs. Collecting in the lakes and streams, he began to notice the conditions in which different species seemed to thrive, and he also puzzled over why some aquatic habitats fostered a great diversity of species and others not so many. Niche theory and the diversity question would become major elements in his thinking about ecology—that is, ecology in its proper scientific sense, meaning the study of how the organisms and physical conditions of a habitat interact.

In Johannesburg, Hutchinson also began his short-lived marriage with Grace Pickford, a fellow student at Cambridge who was in South Africa studying earthworms. They traveled together to Yale, where Pickford would turn her earthworm research into a Yale doctorate and become a staff scientist first at Yale's new Bingham Oceanographic Laboratory and later, when the two institutions merged, at the Peabody Museum.

Hutchinson thought his new Yale colleagues were too provincial and conservative, and he worried that the American system of assigned reading of textbooks deprived students of his own hunter-gatherer experience of learning by discovery. Having come from Witwatersrand, where continental drift was openly debated, he also seems to have learned the hard way that even mentioning the drift idea in Charles Schuchert's Yale resulted in "a deeply silent response comparable to what might be elicited by a grossly obscene remark at a church supper."[3]

But the Peabody Museum "began to play a great part in my life as soon as I stepped into it," Hutchinson later recalled. He was impressed first by the Foucault pendulum then hanging in the lobby, which made the abstract idea of the Earth's rotation suddenly visible, and next by the theatricality of the *Brontosaurus* skeleton in midconstruction. (Many years later, when it was fully assembled, he was delighted by the equally theatrical connection—and the contrast—on seeing the dinosaur's "basic tetrapod structure repeated in the small, living, and moving feminine human body" of a belly dancer performing in the Great Hall of Dinosaurs at a dinner marking the opening of a Tutankhamun show.) But the real treasure for Hutchinson was the Peabody's rich collection of specimens of all kinds, from those of Silliman onward.[4]

"One of the things that truly excited him about coming to Yale in 1928," according to conservation biologist Tom Lovejoy, a later Hutchinson student, was the Peabody display of O. C. Marsh's toothed birds *Ichthyornis* and *Hesperornis,* still among the most potent proofs of Darwin's evolutionary theories. Other specimens thrilled Hutchinson on entirely untheoretical grounds. One of his most popular lectures arose from the experience of standing in the Peabody bird collection: "Suddenly it dawned on me that I had never realized what an extraordinary number of pigeons are bright green. . . . To me, this realization, though it had no apparent value in relation to anything else I knew, gave me intense pleasure that I can still recall and re-experience. Feelings of this sort mold our lives, I think always enriching them."[5]

Hutchinson worked closely with Peabody curators. He donated his water bug collection as well as many of the specimens on which he based his landmark studies to the museum. He also made frequent use of Peabody specimens in his teaching so his students could experience the vivid feelings that come with handling specimens. His first official connection with the Peabody came in 1932, when he traveled to the Tibetan Plateau with the Yale North India Expedition. His job there, at least as the expedition geologist saw it, was to fill his collecting bottles with lots of "queer water animals."[6]

In fact, Hutchinson was already employing his own kit of tools for characterizing not just the plant and animal life in a body of water but also the water chemistry, temperature gradients, ultraviolet light intensity, sediment composition, and paleobiology, among other parameters. He was thinking about how these physical conditions shaped the diversity of life, and about comparisons across the three continents where he had so far gathered data. Hutchinson's approach to the natural world was unusual not just because it was so thoroughly quantitative but because it was simultaneously aesthetic and at times religious. He managed this, moreover, even while rowing a balky inflatable boat with low gunwales and stubby little oars across Himalayan lakes at elevations up to seventeen thousand feet. In those debilitating circumstances, he described the experience of observing three tiny crustacean species:

> From the centre of the lake, the dry burnt colours of the shore are lost completely; the snow falls on the enamelled surface, silently strikes it and rests for a moment, then disappears. In the blue water small black spots move leisurely here and there, and still tinier bright red points can be seen skipping among them, as if the turquoise had suddenly liquefied like the miraculous blood of a saint, while into the liquid blue, tiny pieces of its black matrix and minute particles of red ironstone had floated. In the deeper water and along the shore, small pale shrimps skurry about. These three crustacea, black, red, and pale grey, are the only animals that normally live in the lake.

Hutchinson espoused Simone Weil's definition of science as "the study of the beauty of the world" and regarded it as "a means of enlightenment," according to science historian Sharon E. Kingsland. Like art, it "could produce feelings of exaltation that ennobled the human spirit."[7]

Because this sounds so lofty, it's worth noting that Hutchinson's working circumstances were often in need of exaltation—or, anyway, housekeeping. According to Gordon Riley, who joined the Yale faculty after completing his doctorate with Hutchinson, "Evelyn lived in that magnificent, well-ordered mind, which was a good place for him to be, for he was surrounded by chaos. His clothes were shabby, his cars decrepit. Every surface in his office was piled high with books and papers, although he could instantly locate anything he wanted. . . . One day he came to my lab to borrow some Nessler tubes . . . used in visual colorimetry. He stuck a half dozen in his jacket pocket. They went through a hole in his pocket and smashed on the floor."[8]

Remaking Natural History

From the 1930s on, Hutchinson and his graduate students conducted much of their research at Linsley Pond, a twenty-minute drive from the Yale campus in North Branford. It is a modest little "kettle hole" lake, formed by a huge block of ice left behind by a retreating glacier, and not otherwise particularly unusual. Hutchinson made it special. In one experiment, just after World War II, for instance, he and a student set out twenty-four portions of radiophosphorous on the surface of the lake in one of the earliest successful uses of a radioactive tracer in natural history. (Student lore held that Hutchinson swam the material across the lake or, more likely, dropped a vial of the phosphorous into the bottom of the boat and then had to swamp the boat to get the phosphorous into the lake. The resulting paper merely mentions a boat.) Samples at different depths over the following week indicated that the radioactive tracer got taken up by algae and other aquatic plants, and by planktonic animals that ate these phytoplankton. Then it rained down to the bottom in their droppings and was soon released back into the water, to be taken up again by other plankton. This "biogeochemical cycling" helped demonstrate the intimate connection between the physical habitat and the species living within it. The detailed study of Linsley Pond by Hutchinson and his student collaborators would eventually produce some of the most important findings in the rise of modern ecology.

Hutchinson's science gradually developed into a definitive four-volume *Treatise on Limnology.* (His better instincts had initially inclined him to name it "The Natural History of Lakes.") At about the same time, he was publishing a series of articles that would shape all subsequent ecological research. In a 1948 paper, "Circular Causal Systems in Ecology," he asserted the power of the endless feedback loop from the inorganic to the living world and back again. (If that unified view of the world as a living system sounds familiar, it may be because many of Hutchinson's ideas would later turn up in popularized form as part of the James Lovelock–Lynn Margulis "Gaia Hypothesis.") Next, Hutchinson took on the idea of the biological niche, which had become an essential concept for understanding how a species fit into its habitat but was also vague to the point of being meaningless. In a 1957 paper with the underwhelming title "Concluding Remarks" (from the circumstances of its delivery), Hutchinson proposed a way to quantify all the physical and biological conditions that make up the fundamental niche for a species. Traditionalists disliked this quantitative and theoretical approach at first. They worried that biologists would come to characterize a habitat or a niche by compiling a lot of biogeochemical data, without ever having laid eyes on the thing itself. Perhaps to assuage them, Hutchinson chose what was for him at least a highly improbable analogy. He described his "logicomathematical system" as a tool for the ecologist to bring out, like "a vacuum cleaner," only "when a lot of irrelevant litter has accumulated."[9]

In the same article, Hutchinson took on the equally vague and commonplace idea of "competitive exclusion." It had become biological dogma that, when two species depend on the same resource in a habitat, one will inevitably drive out the other. But Hutchinson thought people were too quick to reach for competitive exclusion as an explanation. Properly describing the niche would make it practical, he argued, to distinguish the effects of changes in the niche from those of competitive exclusion. For instance, people commonly regarded the displacement of Britain's native red squirrels by introduced gray squirrels as an example of competitive exclusion. But closer examination had demonstrated that the driving factor was a change in the niche: human persecution, to which the red squirrels were more vulnerable, had multiplied. The influence of this work "on the ecological community was transformational," according to ecologists Melinda D. Smith and David K. Skelly. "Much of the research effort expended by ecologists during the 1960s and 1970s traced back to Hutchinson's insights."[10] Hutchinson's ecological theories were remaking natural history as a true science.

One of the most influential of his insights arose from a small pond near the hilltop church of Santa Rosalia in Palermo, Sicily. Hutchinson was puzzled that the pond contained a large number of water bugs—but all belonging to just two species of the genus *Corixa*. The 1959 paper "Homage to Santa Rosalia" asked why—and, more broadly, why there are so many kinds of animals in the world. Hutchinson saw that one *Corixa* species was about twice the size of the other, and that its breeding season was ending just as that of the other species was getting under way. This led him to propose that seasonality, differences in size, and the effect of size on the resources different species could target all played a role in shaping the diversity of species in a habitat.

Hutchinson continued to pursue these questions in a 1961 paper, "The Paradox of the Plankton," examining how a seemingly unvaried habitat could support an enormous variety of species. His experience measuring every characteristic of the lakes he studied had taught him the importance of variation the human eye might not ordinarily notice—for instance, differences in light intensity, nutrient concentration, or turbulence at different depths, or the way the presence of one species might in effect create a habitat for a symbiotic partner. Two identical-looking lakes

Hutchinson's work on water beetles and his influential paper "Homage to Santa Rosalia" helped launch modern ecology.

could thus differ dramatically in terms of microhabitats. Moreover, these conditions might change from year to year; equilibrium and the "balance of nature" were not necessarily all that common.

Intellectual Progeny

Despite the accolades this body of work brought to Hutchinson—Stephen Jay Gould, for instance, called him "the world's greatest ecologist"—Hutchinson regarded himself first and last as a teacher, like his father. On receiving one of the many honors and awards that came to him in late life, he dismissed the idea that he was the father of ecology, attributing that title instead to Charles Darwin. "I will, however, proudly admit," he added, "to being the intellectual father of a great number of ecologists, for an extraordinary succession of incomparable young men and women who have studied with me, and I with them."[11]

Among them was Raymond Lindeman, whose 1942 paper "The Trophic-Dynamic Aspect of Ecology" was roundly rejected at first by peer reviewers for the journal *Ecology*. The reviewers were shocked by Lindeman's readiness to generalize based on a detailed study of a single lake. They objected, for instance, to his assertion that food chains are short because of what's become known as the "ten percent law": only about 10 percent of the energy available to plants gets transferred up to the herbivores feeding on those plants, and only 10 percent of that gets transferred up to the predators feeding on the herbivores. The reviewers stood fast even after Hutchinson mounted a spirited defense of Lindeman's work. It was a classic case of radical new ideas running up against "firm resistance from the scientific establishment," according to British ecologist Martin Kent, "yet the work was saved by the willingness of a young editor to over-rule the recommendations of journal referees many years his senior." Lindeman died of a rare liver disorder at the age of twenty-seven before the article could appear in print but after that editor had accepted it. His pioneering statement of the ecosystem ecology concept, together with "the associated concepts of trophic structure, energy flows and pathways," Kent wrote, "have assumed the status of the central paradigm of ecology and many would argue of biogeography as well."[12]

Robert MacArthur, one of the founders of evolutionary ecology, was also among Hutchinson's students. He introduced central concepts of population ecology and later went on to write, with E. O. Wilson, *The Theory of Island Biogeography.* That book's analysis of the geographic distribution and extinction of species provided some of the fundamental thinking for the new field of conservation biology and for the science-based planning of protected areas. Hutchinson also taught many women at a time when they were generally unwelcome in science, among them such prominent names in their fields as Alison Jolly and Donna Haraway.

Incredibly, given this record of success, molecular biologists at Yale in the mid-1960s began systematically to purge Hutchinson's students—and ecology itself—from the faculty. They considered this line of research mere natural history, the stuff of Victorian bird egg and butterfly collectors, utterly obsolete in the DNA era. (It persisted only at the Yale School of Forestry, which became the School of Forestry and Environmental Studies.) The study of the organisms and ecosystems of the living world would not return to the academic mainstream at Yale until the university belatedly established a department of ecology and evolutionary biology in 1997, six years after Hutchinson's death. By then, though, the ranks of Hutchinson's intellectual progeny, many of whom had never met the man except through the power of his ideas, had spread well beyond Yale to include countless researchers on campuses and in conservation groups around the world. Gerardo Budowski, the first director general of the International Union for Conservation of Nature (IUCN), took a single course with Hutchinson at Yale in the early 1960s. But he might have been speaking for the entire conservation and ecology community when he remarked that, even decades later, "My mind is still full of him."[13]

CHAPTER 20

The Beauty of the Beasts

It was like giving birth to an elephant.

RUDOLPH F. ZALLINGER on his *Age of Reptiles* mural

At the time the Peabody Museum was opening its new building in 1925, the major figure in redirecting its course for the rest of the century was at a difficult and demeaning place in his career, emptying spittoons, washing windows, and feeding fish for $110 a month at the New York Aquarium. Albert E. Parr, then twenty-six, had trained as a zoologist and oceanographer at the Royal University of Oslo in his native Norway and then tramped the world for a while with the Norwegian Merchant Marines. Arriving as a new immigrant and newly married in New York City, he had taken the most promising job he could find.

It was an unexpectedly brilliant choice. Harry Payne Bingham, a Yale alumnus with a huge family fortune and a passion for marine biology, soon befriended him. Bingham had already conducted two deep-sea expeditions, the first to the Caribbean aboard his yacht *Pawnee,* the second to the Gulf of California aboard the new *Pawnee II,* specifically designed for deep-sea trawling and research. Parr was aboard for the third expedition, to the Bahamas and Bermuda, in 1927. That same year, he became curator of Bingham's private collection. When the collection moved to the Peabody Museum on loan in 1928, Parr went with it as an assistant curator of zoology. And when the loan evolved in 1930 into the Peabody-affiliated Bingham Oceanographic Laboratory, he stayed on as its director.

Parr took up where A. E. Verrill had left off, conducting annual voyages aboard the Woods Hole Oceanographic Institution's *Atlantis,* a 142-foot-long, steel-hulled, ketch-rigged sailing vessel. But where Verrill had conducted his fieldwork in the summers, Parr often went to sea for months at a time in midwinter. Especially after the comforts of *Pawnee II,* he was appalled by conditions aboard *Atlantis,* among them meat that glowed in the dark, the head in a gag-inducing state of filth, and a sullen, uncooperative, and occasionally violent crew. (They beat up the cook.) Parr nonetheless collected prodigiously, in one case experimenting with a huge new net that yielded, in a single haul, 491 fish of forty-seven different varieties, twelve of them new to science.[1] He also published often and launched the *Journal of Marine Research.*

In 1938, just a dozen years after his brief career as a spittoon man, Parr became Yale's first professor of oceanography and director of the Peabody Museum. This was, in a small way, as if the Vatican College of Cardinals were to choose the Dalai Lama as the next pope: Parr had no great interest in paleontology, until then the Peabody's backbone, if not its sole reason for being. He also regarded the heavy emphasis on evolution as a misguided use of display space, not because it was wrong but because it was "a finished issue," already a notch on the belt of demonstrated scientific truth. (In 2007, contemplating the Creation Museum being masterminded by a Young-Earther named Ken Ham, a writer in the *Chronicle of Higher Education* commented: "Dim the lights, cue the diorama of Ham's evangelical anti-Darwin displays, and watch the rapid spinning of Albert Parr in his grave.")[2]

Beyond Paleontology

Parr believed that display space should help museumgoers understand the world around them, from "dust bowls and Japanese beetles" to the rise and fall of fisheries.[3] He thus launched the Peabody's twentieth-century push into the present-day world of zoology and ecology. For the first time in the museum's history, he also treated anthropology as an important scientific discipline. Among the first initiatives Parr undertook on his new job was to move the artifacts of different civilizations down from the third floor and incorporate them into the Hall of Man, thus making human culture a sort of culmination to the evolutionary narrative.

Mildred Porter had painfully demonstrated that the Peabody's evolution-driven approach, with its 21.40-minute trot through the history of the Earth, wasn't fulfilling the museum's intended mission. This was at least partly because the Peabody's staff had been obliged to work "with some desperation" to prepare for the December 1925 reopening, retrieving their specimens from nine years in the broom closet. "It was a rush job," Carl O. Dunbar later acknowledged. Many labels had been typed out on ordinary paper and were "too small to be easily read, even when new," becoming less legible as they faded and yellowed. The building itself also presented issues: fluorescent lighting had become available only in 1938, so the Great Hall of Dinosaurs relied mainly on skylights, which were dependent on the weather and always inadequate. The twenty-six-foot-high walls had been finished in yellow brick, "an unhappy choice," Dunbar wrote, not least because "the hard surface of the brick reflected every sound so that it was difficult to speak to a class . . . or even to carry on a conversation."[4] Small wonder visitors seemed to be in such a hurry.

Parr began his time as director by successfully lobbying the university for a five-year capital improvements budget. He used much of that funding to commission natural history art as part of the museum's new program to engage the attention of visitors (and perhaps as an unspoken corollary to the idea that human culture was intrinsic to the natural history narrative). G. Evelyn Hutchinson would later write, "Today we enter an art gallery expecting to be delighted by the beauty of certain works of man; we enter a natural history museum expecting to be instructed in the workings of nature." Hutchinson thought—and Parr agreed—that "the beauty of the natural world and its relation to human art deserves more consideration than it is customarily given, and deserves such consideration quite specifically in the context of the natural history museum."[5] Instruction was critical, but both men thought that natural history museums should delight, too.

The Wall Needed Sprucing Up

One space in particular caught Parr's attention. The Great Hall was "a dismal, barren cavern, devoid of color," and he meant to do something about it. He asked the Yale School of the Fine Arts, then an undergraduate program, to recommend a mural painter. A professor at the school replied, "You already know him." In fact, Parr had hired a senior at the art school named Rudolph F. Zallinger to illustrate seaweeds for his own research. Seeing how meticulous his work was, Parr began to talk with Zallinger about the larger project. "Albert had this wall in the Museum and a lot of gray bones," Zallinger recalled. "He thought it ought to be spruced up."[6] Parr had $40 a week available to pay an artist because the electrician had gone off to war.

By the time Zallinger began painting the mural, in 1943, Parr had moved on to become director of the American Museum of Natural History. (More grave-spinning, this time for Henry Fairfield Osborn: one of Parr's first moves there was to disband Osborn's prized vertebrate paleontology department, folding it in with

Rudolph F. Zallinger gulped on first looking down the 110 feet of wall that awaited his mural. But he put his faith in "orderly procedure."

other reptiles.) Zallinger remained behind at Yale, twenty-three years old, with an endless span of empty brick wall staring down at him. He knew nothing about dinosaurs. According to one account, he had to look up the word in the dictionary after being offered the job. "I didn't know the front end from the rear end of a dinosaur," Zallinger admitted in a 1980 interview.[7] Peabody scientists—including Carl O. Dunbar, who had become the new director of the museum, paleontology curator G. Edward Lewis, and paleobotanist George R. Wieland—provided a six-month crash course, Introduction to the Age of Reptiles 101a and 101b, as Zallinger began working up preliminary sketches.

Parr's original plan had been for a series of panels depicting some of the fossil dinosaurs displayed on the floor of the Great Hall as they might have looked in life. Instead, Zallinger had the youthful audacity to suggest making the entire wall into a continuous "panorama of time," 110 feet long by 16 feet high, ultimately spanning 300 million years in the saga of life on Earth.[8]

A handful of species would have to stand in for entire epochs. Even so, the artist's need for head-to-toe detail soon went beyond what the paleontologists could say for certain. Ancient life had only occasionally and accidentally been preserved in fossils, and individual animals had seldom been preserved intact, meaning that

The Progress of Man

Zallinger's murals on the walls of the Peabody made his name. But almost by accident, he also made cartoon history with a visual trope that's instantly recognizable and yet almost never attributed to him. It started with an illustration he produced for the 1965 Time-Life book *Early Man.* In a section headlined "The Road to Homo Sapiens," Zallinger depicted a line of proto-apes, apes, and hominids rising from a crouch to a hunch to the tall, upright modern man. The full foldout spread showed fifteen individuals, starting with *Pliopithecus* and ending with *Homo sapiens.* But when folded in, a more simplified version appeared, with just six individuals. It quickly became known as *March of Progress,* from a line in the text, and it went on to become one of the most famous images in the history of scientific illustration, almost as familiar as Leonardo da Vinci's *Vitruvian Man.*

In fact, similar drawings had appeared as far back as T. H. Huxley's 1863 book *Man's Place in Nature.* The drawing also echoed the Peabody Museum's own 1925 display showing a sequence from gibbon to orangutan to chimpanzee to gorilla to man. But after Zallinger, it became a meme. In the half century since *Progress* first appeared, versions of this drawing have turned up, among other improbable places, on the cover of a Doors album, as the emblem of the Leakey Foundation, and as an ad for Guinness, the final stage in primate evolution evidently involving a pint of "the black stuff." Among more recent parodies, one cartoon depicted modern man as a bloated fast-food

Zallinger drew 15 different primates for the *March of Progress* section of the Time-Life book *Early Man.* But an abbreviated version soon became the universal—though inaccurate—symbol of evolution.

customer who evolves into an actual pig. Another, drawn by *The Simpsons* creator Matt Groening, depicted "Neanderslob" evolving into "Homersapien."

But serving the whim of editors at Time-Life didn't work out so well for science. Evolution isn't necessarily about progress, as that Homer Simpson example might suggest. In his 1989 book *Wonderful Life,* paleontologist Stephen Jay Gould fumed that *March of Progress* had become "*the* canonical representation of evolution—the one picture immediately grasped and viscerally understood by all." But it was a "false iconography," he wrote. "Life is a copiously branching bush, continually pruned by the grim reaper of extinction, not a ladder of predictable progress." (According to one of Zallinger's daughters, Lisa David, her father also objected to the linear layout. He had drawn each figure separately and worried that presenting them as a continuous series misrepresented, "from a scientific point of view, how this whole evolution occurred.") Gould added that his own books were "dedicated to debunking this picture of evolution."[a] But the iconography of *March of Progress* had by then become so powerful and pervasive that it had already turned up on the foreign edition covers of four books by, yes, Stephen Jay Gould.

For some, the dinosaur craze that followed was just another of that era's fads. But many young readers became permanently hooked on dinosaurs, and some of them would find their way to Yale and to the Peabody Museum, not just to admire the Zallinger mural (one of them recalled being "moved nearly to tears" on first standing beneath it),[17] but to study under a paleontologist and fellow Zallinger-admirer named John Ostrom. With their help, over the course of the 1960s and 1970s, Ostrom would develop a radically different understanding of dinosaur life—not plodding and stupid but swifter, smarter, more birdlike—launching what became the modern dinosaur renaissance.

Zallinger's greatest achievement may thus have been to depict one paleontological worldview so brilliantly as to inspire its complete undoing.

CHAPTER 21

The Art of Being Invisible

We must liberate our designs so as to provide the greatest possible freedom of thought and movement for those who come to feast their eyes upon our treasures.
ALBERT E. PARR, "Refuge from Other-Direction," 1973

When Carl O. Dunbar was making the circuit of other natural history museums as a young man, he worried about their tendency to display stuffed specimens in orderly rows — passerine birds here, cranes and cranelike birds over there. It was a sort of taxonomic police lineup, of interest mostly to people who were already specialists. Specimens lying on their sides in rows also failed to tell visitors anything about how these animals lived.

The question of how to engage the public more effectively continued to trouble Dunbar and museum administrators elsewhere through the 1930s as movies, radios, cars, and other distractions of an increasingly urban and technological society drew people away from nature. Natural history museums were also losing out in the race for glamour and grant money to laboratory sciences. They were acutely aware of the threat to their public audience, their budgets, and even their survival.

Dioramas — or, as Dunbar initially called them, "habitat groups" — showing lifelike animals in their natural settings seemed like one way to beat the police lineup problem and win back public interest. "After seeing endless cases with stuffed animals the habitat groups are a delight," Dunbar wrote in his notebook on a 1917 visit to the Smithsonian's Natural History building in Washington, D.C., and delight was what the Peabody Museum was aiming for.[1]

The appeal of dioramas may elude modern visitors at first glance. Jaded and smartphone-addled, we expect special effects and instantaneous answers almost everywhere. The idea that dioramas are "static" or "boring" tends to arise—to the extent that anyone talks about dioramas in the first place. Even in the heyday of dioramas, some researchers objected to them as a diversion of funds from scientific work. At the American Museum of Natural History, the anthropologist Franz Boas thought that the large expenditure on dioramas "armored the Museum, like a dinosaur, against change." But donors relished the apparent permanence of dioramas, and the museum's director, Henry Fairfield Osborn, thought that being able to see animals in their natural habitat would enlighten visitors and help "preserve forever the vanishing wildlife of the world."[2]

A Rage for Dioramas

The word *diorama* comes from the ancient Greek *di,* for "through," and *ő́ρāμă,* a "sight" or "spectacle." Louis Daguerre, better known as a pioneering inventor of photography, coined the word when he set up a diorama theater beside his studio in the 1820s. It featured painted screens that could be lit, like a theatrical scrim, from different angles and from front or back to reveal the changing aspects of a scene. Physical objects in the foreground helped bring viewers into the scene, whether it was the interior of a cathedral or a landscape with a volcano smoking in the distance. Audiences packed the house and were moved to awe by the subtle changes, from day to night, produced by the combination of art and sophisticated lighting.

Natural history museums borrowed the term but experimented with very different kinds of dioramas. The painter and naturalist Charles Willson Peale tried something like the modern diorama early in the nineteenth century when he presented animal specimens against painted backgrounds at his Philadelphia Museum. But his idea failed to catch on, and the museum itself was eventually purchased and broken up by P. T. Barnum. In 1869, the American Museum of Natural History acquired the sensationalistic diorama *Arab Courier Attacked by Lions* by the French naturalist and specimen dealer Jules Verreaux. But that diorama proved too titillating for scientific tastes (the label advised onlookers to imagine "a vibrating roar from the big cat mingled with the bellowing groans of the terror-stricken" camel).[3] Martha Maxwell, a Colorado taxidermist and museum keeper, created a different sort of sensation at the 1876 Centennial Exposition in Philadelphia with dioramas in settings that included running water. Soon after, the American Museum and the Smithsonian's National Museum of Natural History began hiring taxidermists to create scenes of animal groups in their natural settings and in their characteristic poses. A rage for dioramas soon took hold of natural history museums, lasting through the mid-twentieth century.

The Peabody Museum had installed some small dioramas in the years after its 1925 reopening, but they generally lacked the scale and professionalism of dioramas in larger museums. When Parr decided in 1938 to move the anthropology displays down to the first floor, that seemed like an opportunity to catch up. Parr suggested that a series of new dioramas in the vacated third-floor space should focus on the landscapes of New England itself, not someplace at the other end of the Earth. He wanted visitors to understand the wildlife in their own backyards. After Dunbar took Parr's place, he followed through on this plan, turning to the American Museum of Natural History for advice. As he was touring the extensive collection of dioramas there, a staffer asked who would be painting the backgrounds for the planned dioramas, then added, "Would you like to have James Perry Wilson?" The war effort had delayed a planned project, and Wilson was likely to be idled for a time. Wilson expressed his interest on the spot, and Dunbar returned to New Haven giddy with the expectation of hiring "one of the two greatest living background artists."[4]

The Art of the Diorama

First, though, the Yale Corporation had to approve a budget for the diorama. Yale president Charles Seymour was unenthusiastic. But an improbable friend came to the museum's rescue. Wilmarth Lewis, a senior fellow of the corporation, was an Anglophile and an aesthete who had devoted his life and his substantial inheritance to collecting. He went at it with the same megalomaniacal intensity O. C. Marsh brought to his dinosaurs. But Lewis's passion was for everything written by or associated with the eighteenth-century man of letters Horace Walpole.

Mr. Lewis, as he was known to almost everyone, had been nicknamed Lefty by his undergraduate classmates after "Lefty Louie" Rosenberg, a contract killer of that era with New York's Lenox Avenue Gang. The morning of the corporation meeting, Seymour took Lewis aside and said, "Lefty, I don't think the Corporation will vote $2000 to paint a picture on the Museum wall, do you?" Lewis, who took an interest in campus museums, including the Peabody, replied, "Not if you tell them that way—let me tell them." When the meeting ended, Lefty phoned Dunbar and said, "It's in the bag."[5]

Work soon got under way on a thirty-five-foot-long scene of the Connecticut coast. The formidable challenge was to create a convincing representation of an

James Perry Wilson was so confident in his art that he made no attempt to hide the tie-in between the three-dimensional foreground and the painted background.

actual place, down to the blades of grass that grow there at a particular season and even down to a specific hour of the day—in this case, 11:00 a.m. on June 15—with the animals displaying natural behaviors and the world extending out seamlessly to the horizon, all on a stage that, at the Peabody, was just seven feet deep. It required collaboration among a scientist, a "foreground man," and the background artist. For the *Coastal Region* diorama, Stanley Ball, curator of zoology, laid out a design featuring three habitats—the beach, a salt marsh, and a rural upland. Ralph Morrill, the Peabody's foreground man, worked with preparators at the museum to collect the animal and plant specimens from the wild. (The wild in this case being Milford Point on Long Island Sound for the beach and salt marsh, and possibly Guilford, on the other side of New Haven, for the upland.) As a taxidermist, Morrill had a knack for making "birds look like they were going to leap off their mounts," according to Michael Anderson, who studied under him and now continues his work in an office in the basement of the museum. "They looked absolutely alive."[6]

Wilson, then in his fifties, was the presiding genius, though a genius whose unusual ambition was to become invisible. "Wilson sought an art that concealed his art," Anderson wrote, "so that no aspect of his painting detracted from the illusion of actuality." For the first twenty years of his career, Wilson had worked as an architectural draftsman, producing anonymous renderings of proposed buildings. But he also painted landscapes on the side that showed his obsessive interest in subtle shifts in sky, clouds, and atmosphere, whether in the Grand Canyon or on the fog-shrouded cliffs of Monhegan Island on the coast of Maine. "He delighted in painting the same scene over and over again—at dawn, at noon, at dusk, in rain, in storm, and in full sun," according to a former apprentice.[7] Wilson had come to know a curator at the American Museum of Natural History from a chance encounter at Monhegan. When the architectural market collapsed in the Depression, he moved uptown to the museum and began a new career painting diorama backgrounds, perfecting the art over the next dozen years.

Wilson's quest for invisibility was complicated because the background of a diorama was curved to eliminate the intrusion of obvious corners. He responded with mathematical precision, first visiting the scene he intended to paint and making overlapping panoramic photographs at a height of five feet two inches, corresponding to the average person's viewpoint, sometimes followed by additional overlapping panorama series with the camera angled down and then up. He also made an oil painting on-site to capture colors, light, and mood. Then he applied a grid system to the photographs and followed a mathematical formula to transfer what he had seen—first to the background of a scale model and later to the diorama itself. To keep the drawing accurate across a curved surface, Wilson modified his grid system with what he called "adjusted distorted squares," or "unsquare squares." For all this geometric precision, though, Wilson was sufficiently the free-spirited artist that he preferred to do the actual painting in the nude. Or perhaps he just told people so to ensure his privacy as he worked within his enclosed area. (Zallinger, on the other hand, once created a viewing hole in the wall around his own work area. The curious viewer who tried to get a peek at the work in progress was startled to find another eye staring straight back at him—the image from a mirror Zallinger had positioned on the inside of the hole.)[8]

The Egotism of the Invisible

When Wilson was done with the bulk of his work, he withdrew while Morrill came in to finish the foreground details. Then Wilson returned for the critical final step, "the tie-in"—blending the three-dimensional foreground imperceptibly into the bottom of the two-dimensional background. Other background artists tended to

Preparator Ralph Morrill skillfully constructed the foreground to merge seamlessly with Wilson's background painting for the diorama of the Connecticut shoreline.

disguise the tie-in with strategically placed plants or rocks. But "Wilson just put it right out there," said Anderson. "I think this is where his ego did come out."[9] It was dogma that no background could successfully make the tie-in across a water surface. But at center stage in the *Coastal Region* diorama, the scene passes from a diamondback terrapin turtle walking on a mudflat directly and without obstruction to the surface of a shallow creek, which simultaneously reflects the sky and reveals the bottom. Even the black-crowned night heron poised to strike in the shallows might have been fooled by the verisimilitude of Wilson's water.

The *Coastal Region* was such a success on its completion in October 1946 "that other things came more easily over the next few years," Dunbar wrote, with Lefty Lewis often twisting arms and breaking legs on the Peabody Museum's behalf.[10] In the end, the museum was able to create a total of eleven dioramas, complementing the Hall of Southern New England with a Hall of North American Wildlife. The other great background artist of the day, Francis Lee Jaques (pronounced *JAY-kwees*), came in to paint four of the diorama backgrounds in the 1950s, even as Wilson continued to work. Thus a rivalry that had already been acted out at the American Museum of Natural History was repeated at the Peabody Museum.

This was the one other area in which Wilson's ego may have come out, according to Anderson. Jaques had a more traditionally artistic sensibility, rendering wildlife, and in particular birds, realistically but with a distinctive individual style. He thought Wilson produced backgrounds as "giant kodachromes." But of Jaques, Roger Tory Peterson wrote: "Even in his museum backgrounds, which were intended to deceive the eye, he was not slavishly photographic; he introduced a decorative touch, and one can always spot a Jaques habitat group from those of other artists by the pattern of a cloud or the twist of a branch." This kind of personal touch was anathema to Wilson, though he was too mild and polite to put his disdain into words. The two artists, predictably, disliked each other. At one point, after Jaques had completed work on the *Kaibab Plateau* diorama and gone off to another job, Ralph Morrill found himself dissatisfied with the way the tie-in blended with the reddish dirt in the foreground. Wilson, stepping in to fix it, could not stop himself from also overpainting the entire canyon running up the right side of the diorama.[11]

A Way of Seeing

The status of dioramas today is oddly split. People tend to regard them either as "enduring masterpieces," as one administrator at another museum remarked, or "as a dead zoo located in a dark tunnel—to be either avoided or used as a race track."[12] Some museums have dismantled their dioramas on the theory that they cannot compete with IMAX movies and other more technologically advanced displays or because of criticism that their dioramas have embraced a Great White Hunter perspective. Other museums are building new dioramas or enhancing old ones with audio narrations or with interactive devices designed to make them more engaging.

At the Peabody Museum, the general feeling has been that any such addition would be antithetical to the spirit of the dioramas, which should serve as an escape from the abundant distractions of the outside world and even of the museum itself. They are islands of tranquility, like chapels for the contemplation of nature. Those who love them do so for the darkened room, the illuminated scenes, and the "overwhelming attention to detail" that "closes off distractions," according to one recent analysis of the diorama phenomenon.[13] The same schoolchildren who might bolt through the Peabody Museum's nearby display of Connecticut bird specimens often become fascinated by a scene showing some of those birds as if in life. Do they learn anything in the process?

A Connecticut newspaper reporter had the Peabody Museum in mind when he complained in the 1990s that visitors may admire animal specimens in a diorama "but can learn little about them or about the historical or geographical context of the setting." G. Evelyn Hutchinson, who knew his ecology, thought otherwise. "The design and placing of organisms in the [*Coastal Region*] diorama was so good," he wrote, that he took time in the early 1960s to produce an educational pamphlet on all the "ecological principles easily seen in the Hall of Southern New England natural history." They ranged from "Carnivores Usually Larger and Rarer Than Their Prey" and "Color Often Adaptive" to "Variations of Chipmunks with Zones." One had only to look, which is the essence of natural history.[14]

Winged Things

One New Haven boy who used to visit the Peabody Museum in the 1930s was obsessed with winged insects. "I just devoured the books and learned the scientific names of various insects that were around. I made collections and learned how to mount the bugs, butterflies, and moths," he told an interviewer. His interests eventually shifted to gasoline-powered model airplanes. "And as I got further along, I made all sorts of experimental things—ornithopters, autogyros, helicopters." The ornithopter, he explained, was "a wing-flapping thing. It's like a bird. Instead of a propeller, it uses wing flapping."[15] Later, as an undergraduate at Yale, he learned to fly sailplanes. Later still, in the 1970s, Paul B. MacCready became famous for designing *Gossamer Condor,* the first successful human-powered aircraft, and then *Gossamer Albatross,* which became the first such craft to cross the English Channel.

In 1986, MacCready was back at Yale to describe his recent experience building a working replica of a pterosaur with an eighteen-foot wingspan. One thing he remembered from his childhood, he mentioned on that visit—as if it were a trivial detail—was the image of a dragonfly on the wing over a body of water, depicted on the wall of a diorama at the Peabody Museum.[16]

CHAPTER 22

Into the Unmapped World

It is wonderful to be theoretically relevant, politically aware, and rhetorically inspiring. But first, one must know what one is talking about.

CHARLES FRAKE, *Fine Description*

Early in October 1954, Leopold Pospisil, a Yale graduate student in anthropology, took a floatplane into the highlands of western New Guinea and began a fourteen-month stay in the mapless interior. The plan was to meet and study an uncontacted tribe, and those who knew the country assured him that he would die trying. The highland tribes were ferocious warriors and rumored to be cannibals. For the five-day walk from his drop-off point into literal terra incognita, the nervous Dutch colonial government provided Pospisil with an escort of four local constables and a translator. "I was more afraid of the constables than the tribes," Pospisil recalled. "One of them stumbled and the submachine gun went off and I thought, my God, we could be shot like that." Other deadly opportunities abounded.[1]

Early on, the members of the party were climbing a grassy slope when they looked up to see warriors lining the ridge above and emerging from the brush on both sides. The translator quickly explained to the local headman that Pospisil wanted to learn the language and the culture: "He is neither a missionary nor a government official. He wants to write a book about you." The chief had never seen a book, so Pospisil fished one out of his backpack. "He looked at it," turning it over and riffling the pages. "And that was a break. I remember he made *taku-taku,*" mouth open, thumbnail clicking off his front teeth. "He was amazed." The translator whispered to Pospisil, "It is getting alright." The first chief provided an escort to the next chief's territory, and so on to the next one, until he was accompanied by a small army.[2]

The constables soon went home again, and Pospisil set up housekeeping in the Kamu Valley. He began learning the language of the Kapauku, a highland ethnic group of about forty-five thousand people. "In fourteen days I had a vocabulary of over 2,000 words," he told an interviewer. "In three months I could talk about general things and daily activities, and in five months, I had already started to understand when two Kapauku talked to each other. If you understand when two foreigners talk to each other—not to you, because they slow down and open their mouths and articulate much better for outsiders—you're ready, for anthropological purposes."[3]

Leopold Pospisil with his friend Chief Jokagaibo in New Guinea.

On that first visit and three others that followed, Pospisil lived among the Butekebo clan, numbering roughly six hundred people. His ambition was straightforward: "I would topple all these idiotic theories about so-called primitive people," he said later. He made a quantitative analysis of the economy, reporting in detail that his study subjects were not egalitarian believers in communal ownership, as other anthropologists frequently assumed in their studies, but "primitive capitalists" avidly devoted to private ownership and trade. (On one of Pospisil's later visits, missionary newcomers were attracting congregants by offering tobacco after Sunday service. When the tobacco went away, so did most of the congregants. The villagers explained it to Pospisil: "No tobacco, no hallelujah.")[4] Having trained as a lawyer in Czechoslovakia, Pospisil also conducted a systematic study of the Kapauku legal system, which turned out, even in the absence of written legislation, to be rich with hierarchies, codes of behavior, crimes, and punishments.

In the course of a year, Pospisil witnessed two clan wars, in one case being advised that, on account of his prominence, he would be among the first to die. Another time, Pospisil was traveling with his friend the chief Jokagaibo when war-

During fieldwork in New Guinea in 1954–55, Pospisil collected this dog-tooth necklace and bird hairpin.

riors from a rival clan appeared out of the long grass and began firing arrows. They took cover as the arrows fell around them. Then, when the attack seemed to stop, Pospisil stepped out and in their own language yelled, "*You are pigs! You are dogs!* Send me more arrows for the Peabody Museum."[5] So they did, until they ran out, and the arrows—some carved into simple points, some cut diagonally into blades for shearing off flesh, some with a coil of bark like a spring wrapped around the tip, designed to become stuck in the wound—are now preserved in the Peabody Museum. This collecting technique was not normally taught at Yale. But Pospisil's unorthodox fieldwork was part of a surge that was making Yale at mid-twentieth century one of the greatest centers of anthropology research in the world.

Anthropological Beginnings

The Peabody Museum had been accumulating anthropological material from the start, when O. C. Marsh instructed his collectors in the West to keep an eye out not just for fossils but for artifacts of the North American Indians.[6] He also collected on his own—prodigiously, as always. But it was more like high-grade souvenir hunting than anthropology. Alumni and friends of the museum joined in, with the donations for one year including "a beaded ceremonial coat or shirt from the Misses Terry of New Haven," "2 boxes of antiquities from Egypt," and "a Chil-

The 1867 catalogue lists this Egyptian vase, dated to 300–100 B.C., as object 1 in the museum's anthropology collection.

kat blanket from Alaska, collected by Mrs. Kate Foote Coe in 1886."[7] (The exception to the mere souvenir gathering was the late-nineteenth-century work of Marsh's former student and assistant George Bird Grinnell, who made serious anthropological studies of the Blackfeet, Pawnee, and Cheyenne Indians.)

In 1902, the Peabody Museum established a formal anthropology department and hired George Grant MacCurdy to put the collections in order. MacCurdy traveled widely, conducting his own fieldwork in Iraq, Palestine, and Bulgaria. But like other Peabody curators, he often did his best collecting in the basement of the museum. One day he stumbled across a series of boxes, unopened for decades and largely forgotten. They contained what an art journal of the day subsequently described as the world's largest collection of ancient pottery specimens.[8] Marsh had acquired these relics of Panama's Chiriqui Indians on the side over a twenty-year period but had never had the time to look at them again, much less found the space to display them. MacCurdy noted that, as a young man, Marsh had once made a study of an Indian mound in Newark, Ohio. He could not help wondering, a little forlornly, what might have been possible had Marsh chosen anthropology instead of paleontology as his life's work.

Yale University eventually followed the museum's lead, establishing an academic anthropology department in 1931. Edward Sapir, a pioneer in synthesizing linguistics with anthropology, came from the University of Chicago to serve as

Inventing Connecticut Archaeology

Until the destruction of Fry's Cave in the New Haven suburb of Hamden, archaeology in Connecticut was largely an amateur affair. A few curious farmers and local historians collected arrowheads and stone axes turned up by the plow or eroded out of riverbanks. Haphazard digging by untrained enthusiasts also became common as a result of the excitement aroused by Indian displays at Chicago's 1893 Columbian Exposition.

Fry's Cave was an easy target. It was a dry, sunny shelter under a rock overhang, facing southeast, and comfortable enough that the last of a succession of European and African American inhabitants had moved out only in 1856. Then evidence of earlier inhabitants, possibly dating back to the last Ice Age, began to attract local collectors and outsiders trafficking in Indian artifacts.

Just two miles away at the Peabody Museum, George Grant MacCurdy had recently begun conducting the first Connecticut archaeological survey. But he heard about Fry's Cave only on October 18, 1912, as blasting for trap rock was about to begin there. MacCurdy visited that afternoon. Over the next month, he and two assistants from the museum scrambled to gather the last remaining evidence of Indian habitation—mostly shells and broken bones. Among other curiosities, they found remains of elk, no longer known east of the Rocky Mountains. Then one day, after an unusually large explosion to loosen the trap rock, the foreman of the quarry warned MacCurdy not to go under the overhang. It was good advice: at 10:00 that night, thousands of tons of rock came roaring down in an avalanche that buried the site and spilled out onto nearby pigpens, killing forty-odd pigs.[a]

From MacCurdy's perspective, Fry's Cave had already long since been destroyed. The failure of amateurs to use scientific methods meant there was "absolutely no record as to the relative positions of the various objects in the relic-bearing deposits" and no way of knowing to what period they belonged. What happened at Fry's Cave, he wrote, was "an archeological calamity."[b] MacCurdy went on to devote considerable effort to tracking down local archaeological sites, often in the path of development, and to proselytizing amateur collectors in the gospel of scientific archaeology. He founded the Connecticut Archaeology Society, which became a forum for amateurs and professionals to meet, bicker, and gradually become allies instead of adversaries.

Cornelius Osgood and Froelich Rainey, who succeeded MacCurdy at the Peabody Museum, also bickered. In the 1930s, they undertook digs at Connecticut ar-

Froelich Rainey's notes on archaeological sites in Connecticut, with some artifacts found there.

chaeology sites to give Yale students hands-on experience. But Rainey soon left Yale, largely because Osgood was "an irritating sort of fellow" whose idea of collaboration was that "I did the digging and he the writing."[c] Another Osgood student, Irving "Ben" Rouse, took over the Connecticut work and gradually persuaded amateurs that scientific methods—knowing about mapping techniques, dating methods, the importance of hearth pits—produced better results. Rouse also developed a timeline and a system for classifying Connecticut artifacts.

In the 1950s, the Peabody Museum largely shifted its anthropological emphasis to international sites, handing off Connecticut archaeology to the state and to the University of Connecticut. But in the 1960s, Wilmarth Lewis came to the museum with evidence of Indian artifacts on his own property in Farmington, forty miles north of New Haven. Rouse was skeptical at first but soon realized that the property contained significant material. For about ten years, Yale students conducted systematic investigations there, and the accumulated evidence of scrapers, drills, denticulate tools, and white-tailed deer bones identified the site as a seasonal camp for hunting and butchering—dating back perhaps ten thousand years.

chairman of the new department. But Sapir quickly found himself entangled in the ugliest sort of academic politics. Yale president James R. Angell had almost guaranteed an unhappy reception by carving the new department out of the sociology department without bothering to consult the sociologists. Sapir's training with Columbia University's Franz Boas, who emphasized long-term immersion and the individuality of each culture, also put him at odds with the trend for cross-cultural comparisons and evolutionary thinking at Yale. Finally, Sapir was a Jew, one of only four in a faculty of 569, and many of his colleagues shut him out, socially and otherwise. Sapir died of a heart attack in 1939, at fifty-five, and his successor, George P. Murdock, turned the program back to cross-cultural studies.

At the Peabody Museum, Cornelius Osgood succeeded MacCurdy as curator of anthropology. He was Sapir's friend and former student. But he stayed on for decades after—until 1974—and through his students played an important role in building Yale anthropology into a midcentury powerhouse. One of the first issues Osgood attempted to tackle was the Peabody Museum's perennial lack of space for anthropological exhibits. Together with the museum's director, Albert E. Parr, he lobbied for a separate anthropology center next to the museum.

Surprisingly, given the Depression economy, the Yale Corporation approved a plan in the late 1930s. But war intervened, as had happened to the museum itself in 1917, and in the postwar years "the carefully detailed plan for an anthropology museum at Yale became a fiasco," Osgood wrote. "Costs had skyrocketed. The administration had been transformed. Other departmental and personal interests had sprung up."[9]

Osgood, a notoriously difficult personality, was himself part of the problem. By way of preamble, Keith S. Thomson, a friend and colleague, recalled that "people really respected him. Around the curatorial board, when he spoke, everybody listened. Osgood was also one of the kindest, most interesting and intelligent people. He was not a one-dimensional pain in the ass."[10] He was, however, a multidimensional pain, applying his high standards ferociously and without regard for rank or feelings. He summed up his own life, characteristically, in a 1985 article titled simply "Failures."

"Osgood was tough! *God!*" Pospisil recalled, adding that he was also "excellent." But "his students feared him, his colleagues feared him, I think the president of Yale feared him." The story persists that the university walked away from the anthropology museum project because it did not want to deal with him anymore. Osgood put that decision at number four on his life list of seven failures. He also rated himself a failed field researcher: number one on his list was his long and arduous work among the Athabaskan Indians — "a disaster." He "was not the guy who is about to go into a different culture and make friends," a colleague explained.[11]

Caribbean Digs

Even so, Osgood was brilliant at sending other people into the field for this purpose, especially when the focus was on material culture, meaning archaeology. One of his first students was Irving "Ben" Rouse, who started out in 1930 as a bursary undergraduate mowing lawns and raking leaves on the Yale campus. Then Osgood befriended him and put him to work cataloguing anthropology collections. After earning his doctorate in 1938, Rouse spent his entire career at the Peabody Museum, and his extensive fieldwork from Venezuela up to Cuba won him a reputation as the father of Caribbean archaeology.

On his first excavation in Haiti, Rouse ended up, a colleague later recalled, "sitting in a small Haitian jail with me, while the local police chief and his men dug out a large mound at Meillac in north Haiti, where they thought we were digging

Ben Rouse (*left*), the "father of Caribbean archaeology," at a dig in Puerto Rico in 1962.

for pirate treasure. Perhaps that was a good beginning for a very serious and very academic sort of youngster."[12] In fact, Rouse was mostly digging up fragments of shattered pottery. He went on to do so in Puerto Rico in the 1930s, Cuba in 1941, Trinidad in 1946 and 1969, and Venezuela intermittently from 1946 to 1957. He often collaborated in this work with regional scholars, treating them as intellectual equals and coauthoring books and articles with them long before other North American researchers did so.

Rouse was a museum man at heart, intent on putting things in order, first as an undergraduate studying plant taxonomy (and cataloguing for the Peabody Museum on the side), later as a professional archaeologist. "He believed that classification was knowledge," anthropologist William F. Keegan wrote. "If you could identify cultures and place them in the appropriate boxes of time and space, you would produce a complete culture history. Ben never liked the messiness of anthropology."[13]

As he studied the artifacts coming in, Rouse diagrammed them in time and space, revising and refining as better information became available. He used these diagrams to map out the development of cultures. He thus helped define the two sources of Caribbean immigration: from the Orinoco River area of Venezuela into the Lesser Antilles and from Belize and Mexico's Yucatán Peninsula into the Greater Antilles. He also developed the artifacts he found into detailed portraits of different island cultures, notably in his 1993 book *The Tainos: Rise and Decline of the People Who Greeted Columbus.*

Rouse originally dated pottery fragments and other artifacts using stratigraphy (the depth of the layer in which they were found) and seriation (comparison with similar material excavated elsewhere). But in 1951 Yale established one of the first radiocarbon dating centers in the world, around the corner from the museum in the basement of an old fraternity house on Prospect Street. Edward S. Deevey Jr., an untenured assistant professor of biology studying lake sediments, instigated this development with funding from the Rockefeller Foundation. Deevey described his mentor G. Evelyn Hutchinson as "the chief ratiocinator" of the project, that is, its big thinker.[14] The two ecologists also collaborated to produce the journal for the field, *Radiocarbon,* published at first as a supplement to Benjamin Silliman Sr.'s *American Journal of Science.*

Rouse joined them on the editorial board of *Radiocarbon,* and he found that the new technique sometimes verified his dating of Caribbean artifacts, but it also often pushed objects much further into the past. "When I first came into archeology, we believed that the West Indies had only been inhabited a few centuries before the time of Columbus," Rouse recalled.[15] By the time he was done, Rouse had not only established the entire chronology of Caribbean habitation but had traced its beginnings to seventy-two hundred years ago in Trinidad. Some recent scholars have criticized Rouse for painting too broad a picture of Caribbean cultures. But his work made it possible for the Caribbean nations to recover their own precolonial history. A more reasonable criticism is a variation on the one Yale anthropologists have made from the start: because of Rouse, the Peabody Museum now sits on the largest collection of Caribbean artifacts in the world. It also serves a region with a large population of Caribbean descent. And yet it lacks the space to put any of the material culture of their ancestors on permanent display.

Colossal Heads

Anthropology flourished everywhere in the 1950s and 1960s, in part because of two extraneous factors. First, World War II and the early years of the Cold War had forcibly introduced so many young men (and a few women) to other cultures

tán Peninsula. The Smithsonian's Matthew Stirling had excavated there at a site called San Lorenzo in the 1940s. The results led Stirling to propose a pre-Mayan date for the Olmec civilization, best known for its stone carvings, including so-called colossal heads. The suggestion that the Olmecs came first produced outraged disbelief among Mayanists, who thought, as Coe told an interviewer, "that nobody can be earlier than their beloved Maya."[19] Coe became friends with Stirling and decided to give the site a closer look.

San Lorenzo was a floodplain, laced with rivers but with a large, hill-like plateau built up, thousands of years in the past, by unmechanized laborers moving thousands of tons of fill. There and at other sites in the area, the Olmec managed to import huge boulders from a volcano roughly forty-five miles north. They carved these basalt blocks into massive stone monuments. Previous excavations had turned up numerous "colossal heads" weighing eighteen tons on average. They had become as characteristic of the Olmec culture as the stone figures of Easter Island were for the Rapa Nui. Monument 17, depicting a sneering, flat-nosed ruler from three thousand years in the past, was one such find for Coe.

But his most memorable dig at San Lorenzo, said Coe, began at "this sort of unmarked plainstone stele, a slab coming out of the ground" first identified by Stirling, who had not had time to excavate it. Coe set his team to work digging a small trench, cutting across the ridge and gradually revealing the stratigraphy. Soon the foreman came running to get Coe to come see two stones like cartwheels emerging from the ground. As the excavation continued, the life-size figure of a man began to emerge, minus the head. It was a statue of a ballplayer, half kneeling, and the cartwheels were the shoulders, where moveable arms had once rotated. Later, that excavation revealed an entire line of monuments where the Olmecs had ritually broken up their statues in 900 B.C. before moving to another site nearby. For Coe, the best moment was when Stirling and his wife arrived as this spectacular find unfolded; the trio stood above the excavation and reveled in it together.[20] In all, that excavation produced thirty-five stone monuments, more than double the number previously known from the Olmec civilization. The originals ultimately went to museums in Veracruz and Mexico City, with casts of Monument 17 and the "Kneeling Man" now at the Peabody Museum.

Coe's work there, together with earlier work he had done on the Pacific coast of Guatemala, established a timeline for the rise of the Olmecs, just as Rouse had done for the Caribbean. Radiocarbon dating put the earliest evidence thirty-five hundred years in the past. For Central American prehistory, Coe concluded, "this is the fountainhead of everything we know including the Maya." He called the Olmecs "America's first civilization."[21]

A "Lust for Particulars"

Hal Conklin became known for his "fierce commitment" to recording the details of the cultures he studied in the Philippines.

Hal Conklin had not even bothered to come home from the war. The men in his unit, the 158th Infantry Regiment, also known as the Bushmasters, had fought their way across the Philippine island of Luzon and were preparing to move on to Japan. Then the war ended. "I remember crowds of people, tens of thousands of people" in front of U.S. military headquarters in Manila, Conklin said, "all these people wanting to get home immediately, before Christmas." But the Philippines were "heaven for me, all these languages, all these people, very few of them well studied."[22]

He had learned Malay from a coworker in a sorority house kitchen as a freshman at the University of California at Berkeley before enlisting. That served as his lingua franca until he learned Tagalog, the language of Manila. Then, in 1947, he went out for "three and a half months of one of the most concentrated and rewarding ethnographic experiences I have ever had," learning another language and culture among the Hanunóo people at the southern end of Mindoro Island.[23] When he eventually made his way home to finish college in 1948, he carried the notes that would become his 290-page *Hanunóo-English Vocabulary.* Later, as a graduate student at Yale and as a young professor at Columbia University, he returned to the Hanunóo people three more times for further study. After joining the faculty at Yale, he repeated the process for a study of the Ifugao people of northern Luzon.

Conklin became known as one of the great field anthropologists for what colleagues called "an unusually fierce commitment to detail" and a "lust for particulars." He was a master at the practice of "fine description," as Charles O. Frake, his contemporary as a graduate student at Yale, called it—fine in the sense that it provided such a detailed picture of the cultures he studied. A 1958 paper, "Betel

Among the artifacts Conklin brought back from the Ifugao people: a skirt, a statue of a granary idol, and a gong with a handle made from a human jawbone.

Chewing among the Hanunóo," for instance, covered "the history of the use of betel; the physiology and botany of the chew; the Hanunóo ethnobotanical knowledge, taxonomy, and the terminology underlying the selection of ingredients; the behavioral procedures involved in the construction of a quid," and so on for forty-one pages.[24]

To a layperson—and perhaps to the Hanunóo themselves, at first—this might have seemed like some kind of obsessive madness. There wasn't even any literary or narrative style to the betel-chewing article, Frake acknowledged. Nor was there any attempt at theory. "It is wonderful to be theoretically relevant, politically aware, and rhetorically inspiring," Frake continued. "But first, one must know what one is talking about." Conklin knew. The importance of his close attention to detail, said Frake, was that it made "a forceful political statement about the wisdom and value of the lives of people like the Hanunóo on the far margins of political and economic power in the world today."[25]

In passing on the art of fine description to his many graduate students—teaching them how to see and how to describe what they saw—it helped to have artifacts from the Peabody Museum collection in class. One day in the early 1980s, for instance, graduate students in Conklin's Ethnography class arrived to find a ropelike vine, with its bark stripped off and one end fashioned into a hook, waiting on the seminar table. "It was huge, forty feet long," said Joel Kuipers, now an anthropologist at George Washington University.[26] The assignment for the dozen or so students was to figure out what it was and how to describe it—not at length but as briefly as possible, as for a label or index item.

After they had discussed the basics of appearance for a bit, Conklin explained that farmers in the Ilongot area of Luzon used such vines, which they called *taberuk,* to move from treetop to treetop in a section of forest while pruning away the upper branches to open up the ground below for crops. The class struggled for a while with a way to condense all that. Finally one brave student suggested "tree transfer rope."[27]

"I remember the silence in the room after he said it, and Conklin's narrowed eyes," Kuipers recalled. His criteria for such a description were that it be accurate, economical, precise, and valid for the culture using it. But this wasn't a rope, and "transferring" was probably not what the people in Ilongot thought they were doing with it. The students floundered on, gradually getting at the cultural context for the tree trimming. They learned that there was an English-language word, *pollarding,* for a similar sort of tree pruning traditionally practiced by farmers in Europe. To the class's great relief, everyone agreed it would be a "pollarding vine."[28]

Conklin "was a real pioneer in thinking about the relationship between language and the material world," said Kuipers. "One of the things he recognized was that the process of organizing and classifying the objects that you receive from collections is a linguistic process."[29] The words had to derive from the characteristics of the object, and the choice of words could have unintended consequences. For Conklin, the "pollarding vine" class was a way to bring anthropology out of the realm of abstraction and make it real. It was of course also a lesson that could not have happened without the Peabody Museum.

Conklin's work, both in print and in his collections at the museum, now constitutes a permanent repository of the Hanunóo and the Ifugao cultures. This is of course no substitute for preserving those cultures intact. But the modern world has a way of changing everything. Decades or centuries hence, when the descendants of those cultures want to know what they have lost, they will still be able to learn how their forebears worked, talked, thought, and played though the work of Hal Conklin, much as the Blackfeet, Pawnee, and Cheyenne Indians now retrieve elements of their cultures from the books of George Bird Grinnell.

The Hanunóo acknowledged this work in a way that was perfectly suited to a linguistically oriented anthropologist like Conklin: when a Japanese ethnographer later made a study of this culture, he paid particular attention, like Conklin, to vocabulary, and reported that the Hanunóo language now had a word for the Japanese concept of "things related to knowledge." That word sounded to his ear like *konkirin*.[30]

CHAPTER 23

Zoology in the Time of Geneticists

Fine, we'll pay the penalty but you are going to have to listen to the real experts. Because you are not the experts.

CHARLES G. SIBLEY, *Sunday Times* (London), 1974

In the decades after the war, from the late 1940s into the 1970s, Peabody specialists fanned out across the planet. In the Fayum badlands of Egypt, the mammal paleontologist Elwyn Simons discovered *Aegyptopithecus,* a stem primate from before the apes and monkeys diverged. His student David Pilbeam headed off to India's Siwalik Hills to follow up on hominidlike bones brought back and deposited in the Peabody Museum by G. Edward Lewis, a graduate student on the North India Expedition of 1932. It was another case of collecting in the basement of the museum, and for a time, with the help of Simons and Pilbeam, Lewis's *Ramapithecus* bestrode the stage in a dazzling new life as the earliest known human ancestor. (Ultimately, like other celebrities, it proved to be of somewhat duller stuff.)

A team from the Bingham Oceanographic Laboratory was off the coast of Peru conducting one of the first in-progress studies of El Niño, the broadly influential Pacific warming cycle, and Willard D. Hartman was in Jamaica investigating the relationship between sponges and coral reefs. Closer to home, on Long Island Sound, an oceanographer named Howard L. Sanders bent down one day to examine his haul and came up with an entire new order of primitive invertebrates, the horseshoe shrimps, requiring a complete revision of the evolutionary history of the crustaceans. (The order is Brachypoda, but Sanders named the genus *Hutchinsoniella* after his mentor G. Evelyn Hutchinson.) The evolutionary biologist Keith Thomson made the first careful analysis of an intact coelacanth, the "living fossil," frozen on capture in the Comoro Islands. It was a time of great excitement and gathering perils, especially as the wild ride of life in the 1960s unfolded.

The sense of continual discovery around the Peabody Museum was more surprising because the sciences at Yale still suffered from the suspicion and disdain that James Dwight Dana had first confronted in the 1850s. A century later, Yale had gone decades without bothering to undertake any major construction for science facilities. The resulting lack of laboratory space "at a time when the government was beginning extensively to support science hurt Yale badly," Brooks Mather Kelley wrote in *Yale: A History.* Continuing anti-Semitism also meant that Yale "did not take advantage of the great influx of talented Jews fleeing Hitler's Germany." By the 1950s, only 11.5 percent of undergraduates majored in science, half the level at comparable universities. "Our scientists deserve better," Yale president A. Whitney Griswold declared in 1952. "So does the university of Silliman, Dana, and Gibbs. And so, for that matter, do the security and welfare of the United States." (J. Willard Gibbs was the late nineteenth-century mathematical physicist whom Albert Einstein is said to have called "the greatest mind in American history.")[1]

Griswold set out to rebuild the sciences, with an emphasis on physics and engineering. The Peabody Museum did not figure much in these ambitious plans. When Carl Dunbar retired as the museum's director in 1959, Griswold offered the job to ornithologist S. Dillon Ripley, who was an old friend. But Griswold frankly hoped he would not accept because it was such a dull, stuffy place. Ripley felt otherwise and gladly rose to the offer.[2]

That same year, the museum managed to acquire an additional building, in part through Ripley's prodigious talent for fund-raising. The Bingham Oceanographic and Ornithology Laboratories, located behind the museum, was an undistinguished box in an era when Yale was otherwise employing the great modernist architects of the day. Even the skating rink, just up the street, rated the genius of Eero Saarinen. Of the Bingham, on the other hand, one of its inmates later wrote that a "stupid architect put it all together and should have to be sentenced to live with his mistakes." The O & O, as it was sometimes known, nonetheless meant precious space for marine scientists and ornithologists to do their work. As if to remind people how much this work mattered, Ripley timed the building's dedication ceremony to occur on the centennial of the publication of Darwin's *On the Origin of Species.* "Yale's birds and fishes are now wonderfully housed," G. Evelyn Hutchinson declared at the opening, though he could not help adding, "but its insects await their Maecenas."[3]

Before moving on to become secretary of the Smithsonian Institution, S. Dillon Ripley, here cradling a great hornbill specimen, was a Peabody Museum curator intent on dramatically expanding the ornithology collection.

Ripley was at least the friend of many modern incarnations of Maecenas, and he possessed the delicate blend of charm and pushiness required to win their philanthropic support not just for ornithology or entomology but for the entire museum. If some people at Yale and in the larger community thought that the Peabody Museum lacked excitement, Ripley now meant to prove them wrong.

Flight of the "Worldly Pelican"

S. Dillon Ripley was six feet three inches tall, lean, bald, with an aristocratic bearing and a narrow, somewhat beaky nose. The *New Yorker* thought he looked like Noel Coward, or perhaps like "a rather worldly pelican." He was the great-grandson, on his mother's side, of Sidney Dillon, who had made a fortune building the same Union Pacific Railroad that carried O. C. Marsh and his students into the American West. Enough of that wealth survived to give young S. Dillon Ripley a privileged youth: he summered at his namesake's sprawling home in Litchfield,

Connecticut, saw wildlife as a five-year-old on an uncle's ranch in British Columbia, and later traveled with his mother among the smart set in Europe and in India. He became a serious birder in prep school, where his characteristic penchant for colorful amusements also led him to found the Offal-Eating Club, dedicated to dining on roadkill cooked over an open flame. After graduating from Yale in 1936, he began studying zoology at Columbia University but disliked the gritty atmosphere of Depression-era New York City.[4]

Friends—Ripley always had well-placed friends—soon intervened, inviting him to join an expedition to New Guinea. Ripley promptly agreed, later recounting the eighteen-month voyage in his first (auspiciously titled) book, *The Trail of the Money Bird.* After another expedition, this time to Sumatra, he enrolled at Harvard for his doctorate, then went to work as an assistant curator at the Smithsonian Institution in 1942 while finishing his thesis. He had failed the navy physical on account of being underweight and with a history of malaria. But in Washington, he found his way to the U.S. Office of Strategic Services. The OSS was the wartime intelligence and espionage agency, forerunner of the Central Intelligence Agency. It was then sometimes jokingly called the Oh-So-Social for its tendency to hire the wealthy and well connected. (Yale's "dilettante extraordinaire" Wilmarth Lewis was director of the central information division.)[5] But the work was in earnest.

Ripley was soon based in what is now Sri Lanka as a bureau chief overseeing spies and saboteurs around Southeast Asia—but he always kept an eye out for birds. Once he was shaving in his bungalow, cleaning up for a garden party in progress next door for the British and American brass, when he caught sight of a lesser yellownape, a little woodpecker with an olive-green back and an egg-yellow Mohawk sticking out from the back of its head. As the *New Yorker* recounted the scene in a 1950 profile, this was a bird Ripley needed to collect. He grabbed his shotgun and stepped out into the yard clad only in a bath towel and shaving lather. After the shot, only the shaving lather remained, but he seemed not to notice as he sprang for his prize. Then he looked up to see partygoers, male and female, martinis in hand, watching curiously from across the tea bushes. He joined them soon after, more formally attired, and happily regaled Admiral Lord Louis Mountbatten with the many attractions of *Picus chlorolophus.*[6]

After the war, Wilmarth Lewis steered Ripley to a position at the Peabody Museum on the theory, as the *New Yorker* saw it, that the museum needed someone who spoke the same language as Yale graduates. "Dillon is exceptionally house-broken," Lewis said, "and there was a little discussion about his appointment—you know, a good, solid dinosaur man doesn't generally rub elbows any too comfortably with those who move about in the great world—but it's worked out well." On Ripley's arrival in 1946, the zoology department noted warily that he gave "every promise of continued efficient field work and research."[7]

In fact, Ripley went on to increase the ornithology collection from ten thousand specimens on his arrival to one hundred thousand by 1959. Much of that increase resulted from his assiduous pursuit of collections donated by other birding enthusiasts. But Ripley also made frequent collecting trips to India, where he and his colleague Sálim Ali of the Bombay Natural History Society "redefined the ornithology of the Indian region," according to a 2002 appraisal in the *Auk,* "and in so doing became the acknowledged authorities on this diverse and now threatened avifauna." One result was the ten-volume *Handbook of the Birds of India and Pakistan.*[8]

Ripley also made the first modern collecting expeditions in Nepal at a time when the countryside beyond Kathmandu was still otherwise off limits to outsiders. In New Guinea, he had been happy enough to allow people to believe that he was Ripley of Ripley's Believe It or Not! when it seemed there might be some advantage to be gained from it. In the same fashion, he advanced his case with the prime minister of Nepal for unprecedented access by hinting that his very brief acquaintance with the prime minister of India, Jawaharlal Nehru, was a close friendship capable of yielding diplomatic dividends. When the *New Yorker* profile recounted this bit of bluster, the Peabody's Carl Dunbar worried that it might cause trouble for Ripley back in India. (In fact, Nehru was furious.) "Some of your friends here think the article falls far short of doing justice to you," Dunbar fretted. "One of them, who is not a member of the Museum group, remarked in all seriousness that if he were in your place," he would sue the writer for personal injury.[9]

Ripley might not have enjoyed the trivializing tone of the profile, but he certainly did not sue. Like O. C. Marsh before him, he saw value in reaching the public through the press. Colorful stories, more or less true, were the basic currency for winning public attention, and the occasional piece of questionable publicity was an inevitable risk. The 1948–49 Nepal expedition produced twelve hundred specimens, some sent to the Smithsonian Institution, a cosponsor. But most belonged to the Peabody Museum, including the prize, a species found only in Nepal called the spiny babbler, which had been "lost to science" for more than a century. It was of course, headline worthy: "Ripley Finds Rare Bird in Remote Nepal."[10]

"Remington Is Right"

Until Charles L. Remington joined the Yale faculty in 1948, the Peabody Museum had somehow largely neglected 80 percent of life on Earth—in the form of insect species. Remington, a lepidopterist, set out to correct this spectacular omission, creating an entomology department at the same time Dillon Ripley was expanding the museum's bird collection. Remington operated on a minuscule budget and had to deal with specialists in other areas who begrudged the expansion of what one of them called "'dead storage' for insects."[a] Even so, over the course of his forty-five-year career, Remington would acquire a million insect specimens for the Peabody Museum and build one of the most important entomology collections in North America.

Remington was a formidable figure, tall, sturdy, and "full of vitality," one former student recalled, "from his powerful longshoreman's hands to his bushy eyebrows, framing intense probing eyes that missed nothing." He also stood out for his distinctive attire—big, oval eyeglasses, western-style shirts, and bolo ties—and for his head-

Charles Remington (*left*) created the museum's entomology division—and helped rescue the Beinecke, Yale's rare book library, from book-eating insects.

long pace. He seemed to his students to be living "in his own personal wind tunnel, leaning into an otherwise invisible gale." Routinely during tutorial sessions in Remington's office, another student recalled, "his hand would shoot out, mantislike, to successfully snatch an irritating fly, his eyes and attention never diverted from me."[b] It was unnerving.

A trio of Peabody regulars—Remington, Hutchinson, and the zoologist Willard D. Hartman—"had an astonishingly wide range of knowledge of natural history," said Stan Rachootin, another Remington student, now an evolutionary biologist at Mount Holyoke College. "When presenting theory, one just had to look at their expressions to see if Nature might concur." In case you missed the expression, Remington was happy to be vocal and dogmatic in dissent. The first seminar Rachootin taught as a graduate student at Yale, in 1974, "was on possible implications of punctuated equilibrium for evolutionary theory. Charles stood up at the end of the talk, spoke three syllables ('Bilge water'), and stormed out of the room."[c]

And yet Remington's greatest strength was as a teacher. The words *kind* and *patient* frequently turn up in students' memories of the man, oddly paired with that headlong pace and dogmatic manner.[d] He gave his time freely to anyone with an interest in the insect world, regardless of academic affiliation or lack of it. He was, on the one hand, a friend and intellectual sparring partner of many of the greatest biologists of the twentieth century—Hutchinson, Ernst Mayr, Theodosius Dobzhansky, and E. O. Wilson, among others. The novelist Vladimir Nabokov, who was a professional lepidopterist at Harvard when Remington was a graduate student there after World War II, was also a friend, and the two frequently went "lepping" together. (When Remington cofounded the Lepidopterists' Society, Nabokov, otherwise a confirmed nonjoiner, became a charter member. The "Lep. Soc." now has members in sixty nations.)

On the other hand, when a teenage lepidopterist from suburban New Jersey named Rob Raguso sent a letter out of the blue, Remington not only invited him to visit the Peabody entomology department, he happily took the time to show him around what was by then an extensive research and teaching collection. Among other things, that collection specialized in insects from unique habitats, particularly islands, bogs, and ridges, because of what they might say about biogeography. (Raguso was wowed by the birdwing butterflies, some of the biggest and most colorful insects on Earth, in a "long series showing geographic variation" across Wallacea, the transitional zone between Asia and Australia.)[e]

Remington also collected or acquired insect groups with special theoretical, regional, or historical interest. (The Peabody Museum had samples of Connecticut seventeen-year cicadas from every emergence since 1843, and Remington not only continued the series but arranged to create a ninety-acre cicada preserve in nearby Hamden, Connecticut.) He amassed a special collection of extinct and endangered insects, among them the Xerces blue butterfly, obliterated by urban development on sand dune systems in San Francisco in the 1940s. "It is also fair to say," a former student stated, "that there was no aberration in a moth or butterfly that did not stir the soul of Charles Remington."[f] So he collected gynandromorphs—that is, butterflies with mixed male and female traits—as well as insects that were unusually large or otherwise aberrant.

Raguso was delighted by it all, except the "drawers full of cabbage butterflies, those most boring of species to a collector's eye." Sensing his young visitor's disappointment, Remington "reminded me that *Pieris rapae* provided a valuable natural experiment, because we knew the date of their arrival in North America and we could track changes in their morphology and host usage as they respond to different selective factors away from Europe. Everything he showed me had a message rooted in biological pattern and process: butterflies are a terrific model for understanding evolution." Raguso later attended Yale as an undergraduate and is now a chemical ecologist at Cornell University.[g]

Remington's lecture courses in evolution and in terrestrial arthropods were campus favorites, and he conducted them with infectious enthusiasm and a degree of showmanship. He kept a rare Madagascar orchid growing in a room near his office, and to students it always seemed a little magical that it happened to come into flower at just the moment Remington was ready to lecture on one of the most colorful incidents in the story of plant-insect coevolution: seeing this flower, with its fourteen-inch-long nectar tube, Charles Darwin had predicted in 1862 that there had to be a hawkmoth with a tongue that long to pollinate it. When researchers finally found that moth forty years later, they named the subspecies *praedicta* in Darwin's honor.

Remington was also adept enough to take advantage of props that turned up by chance. Once, he was lecturing an entomology class about the wasp genus *Vespa* when a yellow jacket of the sister genus *Vespula* cruised in through an open window. "Yes, just like that one," Remington remarked. "Thank you very much." The wasp circled the room like a model working the runway, then exited.[h]

Teaching, collecting, and conservation (Remington's third great passion) did not

Remington loved seventeen-year periodical cicadas and helped establish the world's only cicada preserve on ninety acres in nearby Hamden.

do him much good in his academic career or as a scientist. He published relatively few articles and no books, sometimes studying topics at length—island biogeography, for instance—without ever bringing the results together in print. The biology department advised Remington, in the midst of its prolonged effort to rid itself of ecologists, that he would never be named a full professor.[i] But instead of taking the hint to go elsewhere, Remington stuck it out (and, thirty-five years into his career, finally became a full professor). In the process, he achieved results that were arguably more enduring than is true of most traditional academic careers.

Remington's joint appointment in zoology and at the Peabody Museum provided "an intellectually and historically based platform" for dissemination of what was then a new conservation consciousness, according to his former student Lincoln Brower.[j] In the tumultuous year 1968, Remington and Paul Ehrlich, then a visiting professor at Yale from Stanford University, cofounded the group Zero Population Growth (ZPG), in part because they had seen so many favorite habitats rolled under by the relentless advance of sprawling human development. ZPG grew to six hundred thousand members in the first year and helped make population issues an everyday topic of debate. Remington was also active in getting California's Channel Islands, where he had collected, set aside as a protected area and ultimately as a national park. But his larger legacy showed up in the generations of conservation-minded students he sent out into the world.

Among them, for instance, Jane Van Zandt and her then husband Lincoln Brower performed now-classic experiments to work out the nature of mimicry and the use of warning coloration to discourage predators in monarch and viceroy butterflies. In what became known as "the barfing blue jay" experiments, they also demonstrated the importance of milkweed in making monarch butterflies unpalatable to the birds that would otherwise be their major predators.[k] Brower later helped map the monarch butterfly's overwintering grounds in Mexico and fought to protect the habitat from illegal logging.

Another Remington student, Robert Michael Pyle, an author and lepidopterist, founded the Xerces Society in 1971. Though originally focused exclusively on endangered butterflies and moths, Xerces has become the leading American voice working to stop the dramatic decline of all pollinating insects and other invertebrates. "The importance of Remington's encouragement of Xerces in these early years, and of the intellectual and scientific climate at Yale, cannot be overstated," Pyle later wrote. Remington was "the godfather—not only for Xerces, but for the American insect conservation movement in general."[l] Decades later, with the monarch butterfly now proposed for endangered species status and with the continuing decline of pollinators costing global agriculture billions of dollars annually, the work of Remington's students is still front page news.

Remington, who died in 2007, would probably have taken all this in stride—and rapid stride at that. Lincoln Brower recalled a characteristic incident from the days when he and fellow student John Coutsis were dealing with Remington's opinionated and occasionally stubborn manner. "One afternoon I walked into Charles's office; he was not there, but over his desk John had installed a huge Remington typewriter advertisement "REMINGTON IS RIGHT," Brower recalled. "Several of us watched as Charles walked in, paused, looked up, and, with a magnificent smile crossing his face, said, "Yes . . . *Yes.*"[m]

Trial Run

Ripley served as director of the Peabody Museum for just five years, from 1959 to 1964, but they were a bright interlude in the museum's history. Many of the initiatives he would later develop on the national stage as secretary of the Smithsonian Institution had their trial run first in New Haven—among them, the emphasis on public access and public education; the push for expeditions to the less studied parts of the world; and the effort to enlist spouses, alumni, and other friends as museum supporters. (To that end, he and his wife, Mary, created what became the Peabody Museum Associates.) Ripley's time at the Peabody Museum gave him a forum for developing his grand worldview and his sense of mission for museums everywhere.

He wanted the Peabody to be a source of excitement and to be talked about. When the gilt artifacts from the tomb of King Tutankhamun toured the United States in 1962, Ripley arranged for the Peabody to be one of the museums hosting the show. The Tutankhamun Treasures tour was a fund-raiser to preserve archaeological monuments threatened by the Aswan Dam project. At the time, the Peabody Museum also had researchers in the field gathering artifacts and studying species that were about to disappear beneath the huge new man-made Lake

Nasser. (The researchers lived and worked on Nile River houseboats that had formerly served as floating brothels.) To promote the Tutankhamun exhibition, Ripley staged a dinner featuring entertainment by a traditional Egyptian dancer accompanied by a professor on a zither.[11] She was a belly dancer, that is. She was supposedly alarmed by the *Brontosaurus* in the center of the Great Hall of Dinosaurs, which she mistook for an extremely large camel, requiring all of Ripley's charm and diplomatic skill to persuade her to go on with the show. It was a hit. President Griswold was too mortified to attend, but society people, people who did not normally include the Peabody Museum in their party schedule (or their philanthropy), were delighted to be there.

As director, Ripley worked to give the living world of animals and plants equal standing in the museum with paleontology, and to that end, he created a medal to honor distinguished workers in the natural world. He named it the Addison Emery Verrill Medal for the museum's first curator of zoology. It would be presented over the years to such eminent biologists as Theodosius Dobzhansky, Ernst Mayr, Peter Raven, and E. O. Wilson and to the paleontologists George Gaylord Simpson and John Ostrom.

The Verrill Medal eventually went to Ripley himself in 1984 to celebrate his twenty-year tenure as one of the most successful secretaries in the history of the Smithsonian Institution. At the ceremony, Ripley told a story about his time at the Peabody Museum. In 1949, a circus gorilla named Gargantua died, and the Peabody Museum, not having a specimen of the world's largest primate, agreed to accept the skeleton. But a young ornithologist on the staff then had the habit of coming to work with his dog, a springer spaniel named Robbie, who should not have been on the premises: "There was a terrible shock," said Ripley, "when one day Robbie was discovered running through the hall with one of Gargantua's humerus bones in his mouth, retrieving it for me. Not the sort of thing one is supposed to do in the Peabody Museum at all." Then, reflecting on this gaffe with the same grace he had managed in the aftermath of the lesser yellownape incident, Ripley added: "Something about the word 'museum' tends to make people feel very slightly dreary, but this is not a dreary museum and all museums, with my thinking, should be places of life and enjoyment and gaiety and fun because that is what education is all about."[12]

Ripley's efforts to boost zoology at the Peabody had not, however, been all gaiety and fun. From the start of his time as director, he faced pressure from more experimentally oriented scientists who thought the study of wildlife was an irrelevant leftover from the Victorian era. That pressure showed up, for instance, in

Ripley's advice to a young ornithologist, Peter Ames, who was studying the rapidly declining population of ospreys, or fish eagles, on the Connecticut River, a forty-minute drive from New Haven. Working with Barbara and Roger Tory Peterson, Ames was among the first to identify concentrations of the pesticide DDT in eggs that failed to hatch, and in 1959, as a graduate student at Yale, he proposed a doctoral study to determine how DDT might be affecting osprey eggs.

But in May that year, a committee appointed by President Griswold to review the sciences at Yale issued a broadly scathing report declaring, among other criticisms, that the zoology faculty "has real distinction only in ecology" and that among the ecologists, "only Hutchinson can be rated as outstanding." The committee thought botany and zoology should be combined in a single biology department. (The botany chairman nervously remarked that "animals eat plants," causing a member of the review committee to send him Darwin's *Insectivorous Plants* by way of reassurance.) The report also advised that both botany and zoology should be under the thumb of "a new professor distinguished in one of the currently important areas of experimental biology." With this stinging rebuke no doubt in mind, Ripley and ornithology curator Philip Humphrey warned Ames that the biology faculty would regard the DDT study as insufficiently weighty for a doctoral thesis. Ames took their advice and instead studied the syrinxes, or vocal organs, of thrushes, while continuing his DDT research on the side. At about that same moment, the author Rachel Carson was pursuing her own work on the DDT issue, confessing to a friend that, if she did not do so, "I could never again listen happily to a thrush song." (But at least she would know what its syrinx had looked like.) Yale's experimental scientists notwithstanding, the DDT connection would become one of the most important science stories of the late twentieth century, beginning with the publication of Carson's *Silent Spring* three years later.[13]

"The Perfect Man"

In April 1963, Whitney Griswold died suddenly of colon cancer at age fifty-six, having only begun his campaign to upgrade the sciences at Yale. The search for his replacement went on for five months. Ripley was probably not stuffy enough for the job—belly dancers in the Great Hall of Dinosaurs and all that. But he was a plausible candidate, according to University of Chicago historian Neil Harris: "nearing fifty, ambitious, energetic, with a Yale degree and influential classmates, possessing an international scientific reputation, administrative experience, and demonstrated capacity at fund-raising."[14] He was so plausible, in truth, that a

Grace under Pressure

In 1969, Grace Pickford, a researcher at the Peabody Museum, became the first woman ever named a full professor of biology at Yale. The undergraduate college was about to become coeducational. Promoting a few eminent but undertitled women already on staff seemed like an easy first step toward overcoming the shortage of women faculty. But "sheer embarrassment" may also have been a factor, according to an administrator of that era.[a] This belated acknowledgment arrived thirty-eight years into Pickford's career at Yale, and with it came an ultimatum from the biology department chairman: she would teach the introductory biology lecture course or she could retire.

Pickford had done no formal teaching since giving up her classes at Albertus Magnus College twenty years earlier to devote herself full-time to research. She was sixty-six years old, a small, shy, reclusive figure, and decidedly different, with close-cropped gray hair, a gravelly British accent, and a habit of dressing in jacket and tie. Teaching introductory biology would put her before a male undergraduate culture that still suffered from a prep school instinct for meeting difference with cruelty. Clement L. Markert, the chairman of the biology department then, was a developmental biologist intent on purging ecologists from the faculty. He may have assumed Pickford would just fade away. Instead, she stepped up to the lectern.

Her introductory biology lectures were thorough and lively enough. But forty-five years later, the biologist and cultural critic Donna Haraway, then a new teaching assistant in that class, mainly recalled the students who heckled from the back of the room. Haraway herself had not realized until that moment that Pickford was a woman, but "what I knew then, felt deeply, and have never forgotten was my shame and fury at Yale undergraduates who were openly disrespectful to Grace in that class over gender and dress."[b] The lecturer carried on without comment and retired the following year.

Grace E. Pickford had grown up in Bournemouth, England, and studied at the University of Cambridge at a time when women could participate in laboratory classes but not receive a degree for their work. She published her first scientific paper there and cofounded the Biological Tea Club with G. Evelyn Hutchinson, Gregory Bateson, and others. At some point, she heard a talk by Alfred Wegener about his continental drift theory, and after finishing at Cambridge, she went to South Africa on a fellowship to research the similarities between earthworms there and in South America. Hutchinson, her new husband, was also working in South Africa, and she collaborated with him on

Grace Pickford, never afraid to get her hands dirty, was the first woman named full professor of biology at Yale. Her gender and style made her a target for some on campus.

his early research into lakes. When Hutchinson received a job offer from Yale in 1928, Pickford followed and earned her doctorate there in 1931, describing fifty new earthworm species. Yale then made her a research fellow at the Bingham Oceanographic Laboratory, affiliated with the Peabody Museum, and she stayed there after her marriage ended in 1934.

At first, she worked with the Peabody Museum collection of squid, octopus, and other cephalopods built up by Addison E. Verrill. She studied *Vampyroteuthis infernalis* (the "vampire squid from hell") and determined that it was so different from both squids and octopods that it belonged in its own order, Vampyromorpha. (By way of perspective, all primates from tarsiers to humans constitute an order. So does the vampire squid.)

She also became a specialist on endocrine function in fish and other marine organisms, often improvising new tools for the kind of microanalytical studies this research required. When pediatricians developed techniques for analyzing the blood of premature babies, for instance, Pickford downsized the technique for studying pituitary hormones in the blood of fish. During World War II, she used her endocrinology background to test the idea that common killifish might become an important protein source. After the war, when field research was still largely off limits to women, a Danish colleague invited her to spend three months at sea with the 1951 *Galathea* expedition in the Indo-Malay region. She was thrilled to be able to work with live specimens, including a vampire squid, and with creatures brought up from six miles deep, "twice the depth at which any living creature has been captured previously." At one point after crew members had taken samples from a large haul of squid, she stopped them from shoveling the rest back overboard again, crying, "*Hi!* You've got at least five different species there!"[c]

Almost from its beginning in 1950, the National Science Foundation supported Pickford's research with a continuing grant. But Yale only grudgingly acknowledged her stature, naming her a lecturer in 1957 and an associate professor in 1959. A year after becoming a full professor in 1969, she retired, accepting a former student's invitation to become the distinguished scientist in residence at Hiram College in Ohio. She took her NSF grant with her, and continued to work and publish there for another decade.

In 1979, Hutchinson published what was effectively his autobiography, *The Kindly Fruits of the Earth*—without mentioning Pickford once. This might have seemed reasonable enough, given that their brief marriage had ended decades earlier. Hutchinson, who had been protecting women on the fringes for his entire life, may have arrived at an impossible choice over which woman to protect. His second wife, to whom he had been married for forty-five years, was then declining with Alzheimer's disease and may have resented his relationship with Pickford. The two of them had remained friends and intellectual partners since Cambridge, and she had regularly made suggestions on his work in progress. She was devastated at being written out of his life. According to Dennis J. Taylor, who was her student then, she sat down in front of the fireplace and one by one tossed each of his letters into the fire.[d]

At Yale today, nine of thirty-eight full professors of biology and about 21 percent of all tenure-track biology faculty are women. The sciences at Yale have also sent Susan Hockfield on to become president of MIT, Judith Rodin to become president of the University of Pennsylvania and the Rockefeller Foundation, and Alison Richard, first female director of the Peabody Museum, to become vice chancellor of the University of Cambridge.

But Grace Pickford was the beginning.

Molecular Hubris

With the discovery of the structure of DNA in the 1950s and the developing picture of the genetic code in the 1960s, molecular biology held the potential to change everything. For zoologists, ecologists, and evolutionary biologists, genetic evidence would eventually become a powerful tool, opening up the story of evolution in ways that had been invisible and even unimaginable before. But their eventual embrace of this new science was delayed for decades, largely on account of academic hubris. Molecular biologists behaved in the beginning as if they had seized the divine fire and they took every opportunity to inform more traditional biologists of their inferiority. Scientists who studied whole organisms, how they lived, and why they mattered were mere "stamp collectors." The naturalists replied in turn that a molecular biologist could spend an entire career studying the genetics of a species without actually being able to recognize the species itself.

It was the beginning of a war that went on in biology departments everywhere over a period of twenty-five years, fought out on the battlegrounds of faculty appointments, budgets, grants, and prestige. The ranks of the dead and defeated were thick with ecologists, evolutionary biologists, taxonomists, systematists, and paleontologists. At Harvard, E. O. Wilson found it prudent to avoid even using words like *ecologist* and *evolutionary biologist.* ("Only later did I sense the anthropological significance of the incident," Wilson wrote. "When one culture sets out to erase another, the first thing its rulers banish is the official use of the native tongue.")[19]

Charles Sibley was the embodiment of this attitude at Yale. "In argument he would bulldoze through, brooking no contradiction," a friend wrote for Sibley's eulogy. "Critics were baited with an acid tongue, and, in fits of temper, he could be a cruel mimic. In short, lesser mortals were not tolerated easily and . . . collegiate friends were few." The eulogist saw nothing "malicious or vindictive" in this. Sibley was just "a big, up-front Yank possessed by 'the big picture' in avian phylogeny and convinced of the righteousness of his cause and invincibility of his intellect."[20]

Those who worked with him saw it differently: "Basically he was a megalomaniac, insecure character with a huge amount of ambition and drive, and it didn't always make a nice mixture," said Frederick Sheldon, now director of the Museum of Natural Science at Louisiana State University. Sheldon worked under Sibley as a genetics graduate student in the 1980s and frequently observed him cutting corners and even "correcting" his data to get the results he wanted. One of the lesser indiscretions Sheldon witnessed occurred the day a television reporter arrived to do an interview. Rather than share the attention, Sibley physically shoved his longtime collaborator Jon Ahlquist out of the room. "But he never understood why people got mad at him," said Sheldon.[21]

Sibley wanted to devise a new and more accurate phylogeny, or family tree, of birds, with species arranged according to their genetic affinities. The DNA sequencing techniques that are commonplace today were utterly unknown in the 1960s. Instead, Paul Johnsgard, a graduate student with Sibley at Cornell, hit on the idea that analyzing egg whites might provide a molecular key for distinguishing one species from another. Sibley ran with it, enlisting birders from around the world to send him egg whites, with the aim of acquiring samples from all of the ninety-five hundred or so bird species then known. In a rare gesture of good manners, Sibley always acknowledged receipt, sending a "thank you" note on Peabody Museum letterhead to his collectors.

The work on egg whites ultimately led to other ways of looking at DNA more directly. Sibley would thus become "one of the founders of the new field of molecular systematics"—that is, the use of genetics to study the evolution of species—which would blossom into "one of the most vibrant fields in biology at the close of the twentieth century."[22] He was within grasping distance of his ambition to become a great scientist.

In 1970, almost inexplicably, given Sibley's lack of even the most rudimentary talent for working with other people, Yale made him director of the Peabody Museum. Henry "Sam" Chauncey Jr., then a special aide to Kingman Brewster, recalled Sibley as "one of those people—and there were a number at Yale—who, on the one hand, got along famously in the social circles of Yale and New Haven, and, on the other, did not get along well with subordinates." That fundamentally snobbish style had been a problem at Yale dating back to O. C. Marsh and even Benjamin Silliman Sr. Chauncey thought Brewster may never even have had occasion to encounter the many nastier sides of Sibley's personality.[23]

Sibley went on to entangle the Peabody Museum in the worst scandal in its history. A 1972 British investigation of the illegal egg trade led to the arrest of a collector, and a search of his home quickly uncovered compromising correspondence: Sibley's "thank you" notes, clearly indicating that he had received egg whites from peregrine falcons, then an endangered species, and from other protected birds. (The peregrine material was a cause of special delight for Sibley. "Needless to say," he assured the collector, "I will make no further comment concerning the source of the specimen!") The British investigators alerted U.S. authorities, leading to a raid by federal agents on the office of the director of the Peabody Museum.[24] Nothing in the Bone Wars between Cope and Marsh could quite match that.

Working with eggs like this one from the lesser bird-of-paradise, Charles Sibley helped pioneer genetic methods to study species evolution. But his disdain for wildlife conservation laws ultimately embroiled the museum in scandal.

great discovery back home? Or should he come out with the truth and risk that the museum would lock up these suddenly precious slabs of rock? Being a mild, "squeaking honest" man, in the words of a former student, Ostrom confessed his belief that it was *Archaeopteryx*.[4]

The curator immediately took back TM6928 and 29 and hurried out of the room. Ostrom slumped in his seat, despairing. A few minutes later, the curator returned with a shoebox tied up with string. He handed it to Ostrom, beaming, and declared, "You have made our museum famous."[5] It was the beginning of something far bigger than either man could have guessed.

The "Dinosaur Renaissance"

In the public imagination then, dinosaurs were plodding, thunderous monsters, cold-blooded and stupid. O. C. Marsh himself had encouraged this notion. He once named a species *Morosaurus,* meaning "moron or stupid lizard," and he thought that *Stegosaurus* had, proportionally, "the smallest brain of any known land vertebrate." The brain size of *Brontosaurus* likewise indicated "a stupid, slow moving reptile." By mid-twentieth century, the tail-dragging dinosaur had become a one-word synonym for failure. Even paleontologists had lost interest in these "symbols of obsolescence and hulking inefficiency," Ostrom's student Robert T. Bakker later wrote. "They did not appear to merit much serious study because they did not seem to go anywhere: no modern vertebrate groups were descended from them."[6]

But dinosaurs had begun to look a lot more interesting as a result of the discovery of *Deinonychus.* Ostrom and his crew spent two full field seasons digging at the hill he later came to call "the Shrine" in Bridger, Montana. (His students called it "Marsh's Nipple.") He spent another three years in study and reconstruction at the Peabody Museum, working with more than a thousand bones from at least four individuals of the same species. Then, in 1969, Ostrom announced what he called a "grandiose" conclusion: The foot of *Deinonychus* was "perhaps the most revealing bit of anatomical evidence" in decades about how dinosaurs really behaved. In place of the plodding, cold-blooded dinosaur stereotype, *Deinonychus* "must have been a fleet-footed, highly predaceous, extremely agile and very active animal, sensitive to many stimuli and quick in its responses," Ostrom wrote.[7]

The "dinosaur renaissance," as Robert Bakker later dubbed it, had begun.[8]

Bakker had been a student member of the 1964 Montana expedition, and he contributed a drawing to Ostrom's paper showing a fleshed-out *Deinonychus* in

John Ostrom (*center*) and his Wyoming field crew in 1962.

full sprint. The "terrible claw" on the hind legs was lifted up and out of the way of its feet, keeping it sharp. That drawing would soon become the icon of the new dinosaur.

Bakker had latched onto many of Ostrom's ideas as an undergraduate and, to Ostrom's occasional chagrin, he treated them as his own. Bakker—"the infamous Bob Bakker," as Peter Dodson, another former Ostrom student, put it—became the outspoken advocate of dinosaurs as active, warm-blooded, and even "superior" animals. "Where John was cautious, Bob was evangelical," Dodson and Philip Gingerich (also an Ostrom student) later recalled. "Each deserves considerable credit for revolutionizing our concept of dinosaurs." In his book *The Riddle of the Dinosaur,* science writer John Noble Wilford added that Bakker "was the young Turk whose views could be dismissed by established paleontologists. Ostrom, however, could not be ignored."[9]

Late in 1969, Ostrom took the challenge directly to the North American Paleontological Convention in Chicago, declaring in a speech that there was "impressive, if not compelling" evidence "that many different kinds of ancient reptiles were characterized by mammalian or avian levels of metabolism." Traditionalists in the audience responded, Bakker later recalled, with "shrieks of horror." Their dusty museum pieces were threatening to come to life as real animals.[10]

Ostrom killed the slow-and-stupid image of dinosaurs, reinventing them as equal in agility to mammals or birds—a transformation that Robert Bakker's drawing of *Deinonychus* made vivid.

Ostrom went on over the next half dozen years to draw out the similarities between *Deinonychus* and *Archaeopteryx* specimens, including the Teylers fossil. Among the titles in a series of landmark papers he published are "The Ancestry of Birds," "The Origin of Birds," and "*Archaeopteryx* and the Origin of Flight." The idea that birds had evolved from dinosaurs was not entirely new, as Ostrom pointed out. In 1836, the Amherst geologist Edward Hitchcock had interpreted dinosaur tracks in the Connecticut River Valley as "a bird track truly, though of giant bulk."[11] And in 1868 T. H. Huxley had argued that the reptilelike bird *Archaeopteryx* and the birdlike reptile *Compsognathus* were closely linked to dinosaurs. But subsequent scientific opinion had shifted to the counterargument—that the many similarities between birds and dinosaurs were merely instances of convergent evolution. Paleontologist Gerhard Heilmann's 1926 book *The Origin of Birds* killed off the dinosaur connection seemingly forever. His argument came down to wishbones: birds had what is properly known as a furcula; dinosaurs didn't even seem to have the collarbones, or clavicles, that fuse to form the furcula.

Ostrom pointed out that this was "negative evidence only and thus inconclusive." A lot of biological features are missing from the fossil record, he noted, but that doesn't make them any less real. Collarbones in particular are delicate and might easily not have survived, or not in any recognizable form. In any case, several dinosaurs with clavicles had already turned up before Heilmann published his book, though no one recognized them as such then. Many more have come to light since then. Against this false negative, Ostrom laid out the positive evidence, listing more than twenty anatomical similarities between *Archaeopteryx* and various dinosaurs.[12] It wasn't just that Ostrom could not be ignored. He was far too thorough and meticulous, and for thirty years too persistent in the face of his critics, to be refuted.

Though one or two holdouts still resist the idea, it is now widely accepted that birds evolved from the group of bipedal theropod dinosaurs that includes *Deinonychus, Velociraptor,* and other familiar species. They are now all grouped together as Maniraptorans, or "hand-snatchers," a taxonomic classification named by Jacques Gauthier at Yale in 1986. When *Nature* published a DNA-based phylogeny of birds in 2015, Yale's Richard O. Prum and his coauthors began by describing the evolutionary history of living birds flatly as "the greatest unresolved challenge in dinosaur systematics."[13]

Feathers

Ostrom had given up fieldwork in the 1970s because of a physical ailment. But he had always believed, like his predecessor Richard Lull, that "the best discoveries are made in museum storerooms," and his own correct identification of TM6928 and 29 had been "a classic example of why a paleontologist or museum should not throw things away that cannot be absolutely classified as worthless."[14] Holed up in his paper-stacked office at the Peabody, studying fossils with his magnifying spectacles, he continued to publish important articles on dinosaurs, birds, and the evolution of animal flight.

As Ostrom quietly continued his work, the dinosaur renaissance spread out from that office to become a "dinosaurian flooding of popular consciousness," as the paleontologist Stephen Jay Gould later put it, with countless books, endless computer-generated dinosaurs on television, and multiple iterations of "Jurassic Park," the last of these directly inspired by Ostrom and *Deinonychus.*[15] The dinosaur renaissance also caused a debate that persists even now among bemused paleontologists and parents alike: why are so many of our children caught up in the raging, teeth-baring grip of full-on dinomania? (The short answer: John Ostrom.)

Ostrom lived to see his ideas about the dinosaur origin of birds—and the feathered plumage of dinosaurs—vindicated by a series of remarkable fossils from northeastern China. It began in 1996 with *Sinosauropteryx* ("the Chinese flying lizard"), a small theropod dinosaur with a mantle of short, dark, featherlike filaments on its back. ("I literally got weak in the knees when I first saw photos," Ostrom told a reporter.)[16] And it culminated in 2012 with the thirty-one-hundred-

Some dinosaur traditionalists have resisted the idea. But it is now clear that many dinosaurs, like *Deinonychus,* were feathered.

pound *Yutyrannus,* a *T. rex* relative only bigger, fiercer, and covered in tufts of long, filamentous feathers. That same year, Canadian researchers browsing through a museum collection of amber found actual dinosaur feathers and "dino fuzz." The *Jurassic Park* movie franchise has thus far continued to hold out for scaly, reptilian dinosaurs, to the chagrin of better-informed children in the audience. But scientific and popular renderings of many dinosaur species now look as ornately feathered as faux Indians at Mardi Gras.

On Ostrom's death in 2005, aged seventy-seven, from a series of small strokes, the *Los Angeles Times* wrote that he had "almost single-handedly convinced the scientific community that birds are descended from dinosaurs." "John Ostrom," the *Sunday Times* (London) added, "did more than anyone else to make dinosaurs interesting, real and visceral." NPR's *All Things Considered* marked the occasion by interviewing Ostrom's first research student, Robert Bakker, and for a moment, recalling the fractious relationship between these two allies in the dinosaur renaissance, the entire paleontological world held its breath. Asked how important Ostrom had been to dinosaur paleontology, Bakker graciously commented: "Nobody was more important."[17]

Eight years before his death, Ostrom himself had already published what was his own best epitaph, according to Daniel Brinkman, Ostrom's last student and now a paleontologist at the Peabody Museum. In "the last word" to his paper "How Bird Flight Might Have Come About," Ostrom wrote: "The missing unknowable fossil record can never be allowed to stifle our curiosity."[18]

There is a postscript. Heading back to his hotel that day in 1970, after having discovered the first new *Archaeopteryx* in fourteen years, Ostrom had to stop at a public restroom. Afterward, he continued on his route, perhaps caught up in contemplating the struggle and the triumphs ahead.

Suddenly, to his horror, he realized that he was empty-handed. The shoebox containing not just TM6928 and 29 but Ostrom's destiny, the fame of both the Teylers and the Peabody Museum, the course of paleontology for decades to come, and the not-yet-imagined dreams of untold armies of dinosaur enthusiasts was perched, abandoned, on a washbasin in a public restroom. Frantically hurrying back, Ostrom found the shoebox untouched. He snatched it up and clutched it to his breast all the way back to the hotel, and to New Haven. Thus John Ostrom saved the future for dinosaurs.[19]

Epilogue

We cannot exist without our museum. The use and utility of its collections is validated time and again and will continue to be so as long as our culture survives.

S. DILLON RIPLEY, dedicating a new wing of the Peabody Museum, 1959

At some point in O. C. Marsh's travels, when he was not doing combat with Cope or basking in the adulation aroused by his work on the toothed birds *Hesperornis* and *Ichthyornis,* his attention fell on an assortment of fossils from the village of Lyme Regis on the southwest coast of England. As was his habit (and very nearly his addiction), he bought them. Among them was a squidlike cephalopod, which took its place on a shelf in the collections of the Peabody Museum and waited, largely unnoticed, for the next 129 years. This is how natural history museums work, and to some outsiders, it might seem like a waste of time and money. But Marsh (however eagerly he may have craved that fossil in the moment) was merely fulfilling his uncle George Peabody's dictum that education is "a debt due from present to future generations."[1]

In the fall of 2006, a graduate student named Jakob Vinther arrived at Yale. He signed up for a class about what causes some fossils, or some features of fossils, to be exceptionally well preserved. Vinther's class project was to investigate why the ink sacs in fossil squids are often preserved in three dimensions, "as solid organic blobs," while every other feature of the animal gets squashed flat. The project was the brainchild of Adolf Seilacher, a paleontologist who was in equal measure notorious and admired for recruiting colleagues to work on his big ideas and then peppering them with hard questions throughout. (Those who experienced this were said to have been "Dolfed.")[2]

This was the Lyme Regis squid's moment, and it came out of storage to be scrutinized under a scanning electron microscope. Other ancient cephalopods followed, with similarly promising results: magnifying the ink sac forty thousand times revealed that it was packed with peculiar little spheres. Vinther recognized them as particles of the color-imparting substance melanin. They survive fossilization so well that early bone hunters sometimes extracted fossilized squid ink, added water, and used it to write letters advertising their finds to potential buyers. Vinther thought that if melanin was preserved in squid fossils, it might also be present in the fossils of other animals as well.[3]

"The whole situation was quite fortunate, that the discovery was made at Yale," said Vinther. His professor in that class was Derek Briggs, "one of the key players in fossil preservation," and just down the hall was Richard Prum, "an expert on bird colors. It was a perfect situation for me to make this discovery because I had these people around, and we could discuss these things, and there was quite a fantastic synergy that developed." The work moved on to a study of melanosomes in birds, the organelles that contain the melanin in feathers, and it became apparent that the size, shape, density, and distribution of different melanosomes corresponded with different colors. Finally, in 2009, the researchers traveled to northeastern China, hoping for permission to examine one of the feathered dinosaur fossils being uncovered there. Chinese paleontologists graciously showed them a recently uncovered dinosaur that had lived in the mid-Jurassic, roughly 160 million years ago. It was, said Vinther, "simply astounding, all the feathers there and beautifully arranged on the body."[4]

What it revealed was even more beautiful than they had imagined. When the researchers examined the melanosomes on *Anchiornis huxleyi,* they were able to identify the original colors feather by feather, and even track color shifts within individual feathers: a red crest like a Mohawk fanned out from the back of *huxleyi*'s head, with red speckles against the gray field of its cheeks. The wings were mainly white but with black epaulets, black along the trailing edges, and black in rows of thickly dappled spots running out to the wingtips. Luxuriant white fringes with black tips feathered out from the backs of its legs. Marsh, Lull, Ostrom and the rest would have been astonished to see the result when it was published in 2010 in the journal *Science:* it was the first time in 66 million years that an entire feathered dinosaur had stood before the world in its true colors.[5] And it looked like a cross between a gaudily costumed Las Vegas showgirl and a spangled Hamburg chicken.

None of those earlier scientists would have been surprised, however, that a neglected fossil preserved in a natural history museum would have played a seminal

Peabody Museum research made *Anchiornis huxleyi* the first feathered dinosaur to stand before the modern world in its true colors.

role in the discovery. This is what natural history museums do: they collect specimens, and over time those specimens begin to tell stories about the planet, stories that otherwise might have remained forever untold.[6]

The Value of Fine Collections

Do those stories still matter in a world that has changed irrevocably over the past 150 years? Will natural history museums still have stories to tell, or the means to tell them, in another 50 or 150 years? When George Peabody founded his museum at Yale in 1866, there were 32 million people in the United States, not 320 million, and just over a billion on the entire planet, rather than the 10 or 12 billion expected to live here later in this century. Humans were an almost universally agricultural species rather than a predominantly urban one. Nature was outside everyone's door, and vast areas of wilderness survived intact. There was no hint of climate change. There was no sixth great extinction. The founding of natural history museums was so popular then because they seemed to celebrate a world that was only becoming more wonderful.

In the present day, the notion has become commonplace that natural history museums are a relic of the nineteenth century, shrouded in the gloom of the vanishing natural world. They are as a result routinely underfunded, strung along in a *danse macabre* with financial ruin. The Academy of Natural Sciences of Philadelphia, the nation's oldest natural history museum, faced severe financial difficulties and had begun to sell off parts of its collection before becoming affiliated with Drexel University in 2011. Other museums have survived only by drastically reducing their scientific staff. They still have their collections, and their public still comes through the door. But they no longer employ enough scientists to interpret those collections adequately for the visiting public or the world at large.

The Peabody Museum is less vulnerable to that kind of reduction because its curators are all primarily Yale University faculty. Their numbers can be reduced only in the unlikely event that the various science departments in which they hold their appointments lose funding or somehow turn their backs on the museum. Yale has thus far followed the logic enunciated by Wilmarth Lewis: "And fine collections attract great scholars, great scholars attract brilliant disciples, the quality of the institution's teaching is improved, and the students are the gainers thereby. Everyone is benefited by collections, whether they know it or not."[7]

Other university museums—including the now-shuttered paleontology museum at Princeton and the Museum of Comparative Zoology at Harvard—have suffered because their central campus location and a lack of adequate parking made them seem off limits to the public. The Peabody Museum, on the other hand, has inadvertently gained from its early twentieth-century move to Science Hill, which locates it as much in the city as on campus, and it is fortunate to have an adequate parking lot next door.

This is not to say that the Peabody Museum has adequate funding. It is the only museum at Yale that must charge admission, "a demographic knife," in the words of one Peabody alumnus, which inadvertently cuts off many area residents who are already cut off by their circumstances from other means of contact with the natural world.[8]

Moreover, many of the museum's public displays remain rooted in the science and design ideas of its 1925 opening and badly need updating. But where the Yale Art Gallery was able to raise $135 million for its renovation completed in 2012, and the Yale Center for British Art is undergoing a yearlong renovation, the Peabody Museum has struggled to raise $30 million to renovate its Great Hall of Dinosaurs. There is, as Lefty Lewis would surely argue, a world of good things to be said for properly displaying J. M. W. Turner seascapes, Indo-Pacific Art, and the furniture masterpieces of colonial Boston and Philadelphia. But surely the story of life on planet Earth has its merits, too.

The New Age of Discovery

The best argument for the continuing relevance of the Peabody and other natural history museums in the twenty-first century is that the stories they tell continue to unfold, and the planet is still worth celebrating. We live in what Yale botanist Michael Donoghue and his coauthor William S. Alverson of the Field Museum described in a 2000 article as "a new age of discovery." Surprising finds not only continue to occur—of both new species and new relationships among species—but do so at a faster rate than in the nineteenth-century heyday of natural history. Nor are these discoveries simply a matter of filling in gaps in past knowledge. On the contrary, many of them are capable of inducing "the sense of awe, amusement, and even befuddlement that remarkable new organisms inspired during the last great age of discovery" and of inspiring among visitors of all ages the kind of wonder a six-year-old experiences on first visiting a natural history museum.[9]

For instance, when field researchers John and Terese Hart, curatorial affiliates of the Peabody Museum, found what appeared to be a new species of monkey in a remote area of the Democratic Republic of Congo, they tracked down specimens from bushmeat hunters and in one case from a predator kill. Genetic and anatomical analysis back at the museum determined that this monkey was different from any known species. The result, the lesula (luh-SOO-la) or *Cercopithecus lomamiensis,* was only the second new primate from Africa to be discovered in the past thirty years. Its long, endearingly lugubrious face helped make it one of the top ten new species of 2012 and launched the *Guardian* (London) into rhapsodies: A photograph of the lesula captured "a sensitivity and intelligence that makes this monkey look like it is sitting for its portrait by Rembrandt. It reveals a staggeringly insightful, wise, and melancholy face. . . . The lesula looks right back at its beholder, calm and pensive, examining you as you examine it." Taking a very different point of view, John Hart commented of lesulas, "They have giant blue backsides. . . . Bright aquamarine buttocks and testicles. What a signal! That aquamarine blue is really a bright color in forest understory."[10] How could anyone not love such a combination of the sublime and the absurd in a fellow primate?

For Eric Sargis, a Peabody anthropology curator who coauthored the scientific description of the lesula, it was a reminder that "major discoveries like this are still possible. . . . And if we're finding new species of primates, then who knows how many new species of small mammals or lizards or insects, just to name a few, might be out there."[11] Nor do these new discoveries occur only in dangerous or remote regions. In 2014, the Peabody's Gregory Watkins-Colwell was part of a team that discovered a new species of frog, the Atlantic Coast leopard frog (*Rana kauffeldi*) in New York City as well as in Connecticut and New Jersey.

Fast, low-cost DNA sequencing now also contributes to the wealth of new discoveries. So does digitization of museum records, by making it easy to detect useful patterns with the help of databases correlating genetic, geographic, morphological, and behavioral details. As late as the 1980s, Peabody Museum staff would ritually log new specimens into clothbound ledgers, inscribing the details with a "crow quill" pen dipped in India ink. The pen wasn't actually crow, or even quill. But it was undeniably Dickensian, and reliance on that sort of record keeping at museums everywhere was a major impediment to knowing where in the world different specimens were located, much less making sense of them.

Starting in the 1990s, the Peabody Museum became a pioneer in the move to digitize natural history museum collections everywhere. A quarter century on, it has 65 percent of specimen lots digitized, whereas most other natural history museums are lucky to have passed the 10 percent mark. The digital record includes information about the collecting event and locality (the Peabody's *Brontosaurus*, for instance, was excavated from "Reed's Quarry 10 at Como Bluff"), photographs of the specimen and label where possible, and increasingly also genetic information. These records are available to anyone via websites, including idigbio.com and gbif.com (the Global Biodiversity Information Facility). That makes the Peabody Museum's 14 million specimens more relevant because other researchers can easily find them and incorporate them into their research.

Digitizing museum collections also helps overcome one of the major roadblocks to species discovery. Most of the world's natural history museums, and most of the preserved specimens, are located in industrialized nations. But most of the undescribed species are in a handful of nations with emerging economies, such as Brazil, South Africa, India, and the Philippines. Ornithologists in Bangalore or Kathmandu might not be able to travel to New Haven to study Dillon Ripley's extensive collection of birds from the Asian subcontinent, but they can at least access the collection via the Internet.

Digitized collections also make it easier for scientists at the Peabody Museum to ask novel questions. According to long-standing theory, for instance, the eyespots on the wing surface in many butterflies evolved to deter or intimidate potential predators, or as signals in sexual communication. But Antónia Monteiro, then a biologist at Yale, wanted to know how those eyespots got there in the first place. She organized a team to photograph every Peabody Museum species in the Nymphalidae, the largest butterfly family. Sophisticated bioinformatics software identified specimens with eyespots and coded the location and size of the spots. Monteiro then overlaid that data onto the evolutionary history of nymphalid butterflies and traced back the origin of eyespots to 90 million years in the past. The ancestral pattern consisted of four or five small eyespots on the underside of the hind wing. It took millions of years of evolution — and many, many butterflies being eaten by predators — for the eyespots to enlarge and migrate to a more useful position on the upper side of the wings.

In a similar "Why is the sky blue?" line of inquiry, Michael Donoghue is currently puzzling over a mystery that is so commonplace most of us never think about it. The leaves of maples, oaks, and other trees in colder climates typically come with serrated edges and strangely angled lobes. But tropical leaves are mostly smooth and unsegmented. Why the difference? Donoghue's hypothesis is that winters in colder climates oblige leaves to spend the early stages of their lives packed tight inside tiny buds no bigger than the nail on your little finger. Working with scanned images of the museum's collection of plant specimens, he is now using automated methods to measure the leaf margins in hundreds of botanical specimens. With the help of CT scans of leaves in the bud, mathematicians are collaborating on the project to determine whether the serrated edges and lobes might allow for more compact folding and packing within the bud.

There is probably no practical benefit to any of this research. There is no likely commercial or medical outcome (though basic research often yields surprising results). But when we stop asking questions like how do leaves grow, or when did butterflies get their spots, we surrender our sense of wonder. We forget how to see or think. We become less human.

At the Peabody Museum, the inclination to ask those questions has persisted, and even increased, and the museum has positioned itself to make the asking easier. Thus the legacy of George Peabody, who supposedly had "no gift except that of making money," seems likely to endure and to continue changing the world long after all of us are gone.

THE PEABODY'S SCIENTISTS

Adams, Phillip A. (1929–98). Visiting curator in entomology specializing in Neuroptera, 1968–69.

Adelberg, Edward A. (1920–2009). Interim curator of entomology and acting director of the museum, 1994–95.

Ague, Jay J. (1959–). Curator of mineralogy, 1998–present, and meteoritics, 2013–present. He was also acting director of the museum in 2008.

Ahlquist, Jon E. (1944–). Curatorial associate in ornithology, 1972–86. He and C. G. Sibley developed an early DNA-based bird phylogeny.

Allen, Robert (?–). Chief preparator for John Ostrom, 1982–94.

Ankel-Simons, Friderun A. (1933–). Curatorial associate in vertebrate paleontology, 1971–77.

Archibald, J. David (1950–). Curator of vertebrate zoology (mammals), 1979–83.

Ashmole, N. Philip (1934–). Associate curator of ornithology, with a focus on island birds, 1968–72.

Ball, Stanley C. (1885–1956). Lab assistant and curator of zoology, 1911–16 and 1926–54.

Baur, George (1859–98). Assistant in osteology specializing in vertebrate paleontology, 1885–90.

Beaman, Reed (1961–). Associate director of informatics, 2003–7.

Beecher, Charles E. (1856–1904). Assistant in paleontology, curator of geology, and trustee of the museum, 1888–1904. He specialized in Paleozoic arthropods and was a leader in trilobite research.

Bennett, Wendell C. (1905–53). Researcher in anthropology specializing in Andean archaeology and Mexican ethnology, 1945–53.

Bertucci, Paola. Curator of historical scientific instruments, 2012–present.

Bostwick, Thomas A. (1857–1923). Assistant to O. C. Marsh, 1872–1923. He made the first papier-mâché mount for Marsh and prepped rhino teeth for R. S. Lull.

Bradley, Frank H. (1838–79). Curator of the Yale natural history collections, 1865–67, and collected fossils for the museum in the summers of 1864 and 1865.

Briggs, Derek E. G. (1950–). Curator of invertebrate paleontology, 2003–present, and director of the museum, 2008–14.

Brinkman, Daniel L. (1960–). Museum assistant and archivist in vertebrate paleontology, 2001–present.

Brown, Charles R. (1958–). Curator of ornithology specializing in behavioral ecology and social behavior, 1986–93.

Brown, Kirby W. (1940–). Curatorial associate in entomology specializing in Coleoptera, 1970–75.

Brush, George J. (1831–1912). One of the three original curators of the museum, in mineralogy, 1867–74. He founded and curated the Brush Mineral Collection until 1904.

Burger, Richard L. (1950–). Curator of anthropology, 1983–present, and director of the museum, 1995–2002.

Burkenroad, Martin D. (1910–86). Assistant curator of the Bingham Oceanographic Collection, 1934–45.

Bush, Katharine J. (1855–1937). Assistant in zoology, 1879–1910. In 1901, she became the first woman to earn a Yale Ph.D. in zoology.

Buss, Leo W. (1953–). Curator of invertebrate zoology specializing in coral reef ecology, 1980–present.

Butts, Susan (1973–). Collections manager of invertebrate paleontology, 2004–present.

Cellinese, Nicoletta. Collections manager of botany, 2002–7.

Challinor, David (1920–2008). Executive assistant and coordinator of exhibits, then deputy director of the museum, specializing in forest ecology, 1961–66. He was also acting director in 1964.

Chandler, Christine (1961–). Museum assistant in vertebrate paleontology, 1988–2000.

Chang, Kwang-chih (1931–2001). Curator of anthropology specializing in Chinese archaeology, 1961–77.

Chatrath, Prithijit S. (1941–). Senior preparator in vertebrate paleontology, 1970–81.

Chinchilla, Oswaldo (1965–). Assistant curator of anthropology, 2014–present.

Clarke, Samuel F. (1851–1928). Assistant in zoology specializing in Hydrozoa, 1874–76.

Cloud, Preston E., Jr. (1912–91). Preparator and assistant in invertebrate paleontology and researcher specializing in brachiopod evolution, 1937–40 and 1940–41.

Coe, Michael D. (1929–). Curator of anthropology specializing in Mayan civilization, 1966–94. He was also acting director of the museum, August–September 1993.

Coe, Wesley R. (1869–1960). Curator of the zoological collections specializing in Nemertea, 1910–27.

Colten, Roger H. (1957–). Collections manager of anthropology, 1997–present.

Conklin, Harold C. (1926–). Curator of anthropology specializing in ethnoscience and linguistics, 1974–96. He was also acting director of the museum, July–August 1992.

Cooper, G. Arthur (1902–2000). Researcher in invertebrate paleontology specializing in Paleozoic Brachiopoda, 1928–30.

Crane, Peter R. (1954–). Curator of paleobotany, 2009–present.

Crompton, Alfred W. (1927–). Curator of vertebrate paleontology and director of the museum, 1964–70.

Daly, Radley H. (1925–2010). Acting director of the museum, July–August 1987.

Dana, Edward S. (1849–1935). Curator of mineralogy, 1874–1922, and trustee of the museum, 1885–1929.

Dana, James D. (1813–95). B. Silliman's assistant and curator of the mineralogy collection, 1836–38, and Chairman of the Board of Trustees of the museum, 1866–95.

Darby, Fred W. (1883–1965). Preparator and chief preparator in vertebrate paleontology, 1907–49. He became a museum guard after his retirement.

Darnell, John C. (1962–). Curator of archaeology, 2011–present.

DaRos White, Maureen P. (1968–). Museum assistant in anthropology, 1993–present.

Deevey, Georgiana B. (1914–82).Researcher at the Bingham Oceanographic Laboratory specializing in plankton, 1949–66.

Delevoryas, Theodore (1929–). Curator of paleobotany and botany, 1962–72, and acting curator of the Herbarium, 1968–69.

deTerra, Hellmut (1900–1981). Researcher in geology, 1931–36, and led the Yale North India Expedition, 1933–34.

Donoghue, Michael J. (1952–). Curator of botany, 2000–present, and director of the museum, 2003–8.

Dove, Michael R. (1949–). Curator of anthropology, 2007–present.

Drew, Daniel J. (1980–). Assistant in invertebrate zoology, 2001–present.

Drew, Leslie C. (1926–2012). Assistant to the director and associate curator in entomology specializing in arachnids, 1967–68.

Dunbar, Carl O. (1891–1979). Curator of invertebrate paleontology specializing in fusuline Foraminifera, 1920–59, and director of the museum, 1942–59.

Duxbury, Alyn C. (1932–). Researcher at the Bingham Oceanographic Laboratory, 1960–64.

Eaton, Daniel C. (1834–95). Founder and curator of the Herbarium specializing in ferns, liverworts, and mosses, and one of the first professors of botany in the United States, 1864–95.

Eaton, George F. (1872–1949). Curator in osteology, 1898–1920, and associate curator of paleontology specializing in ancient reptiles and birds, 1904–20.

Ebeling, Alfred W. (1931–). Assistant curator of ichthyology and researcher at the Bingham Oceanographic Laboratory, 1960–63.

Emerton, James H. (1847–1931). Assistant in zoology, then an assistant to A. E. Verrill and O. C. Marsh, 1880–88. He was also an artist and made illustrations for Verrill and Katharine J. Bush.

Evans, Alexander W. (1868–1959). Curator of the Herbarium, specializing in Marchantiophyta and other bryophytes, 1928–47.

Faller, Eleanor Warren (1948–). Assistant and collections manager of mineralogy, 1974–2009, and collections manager of historical scientific instruments, 1987–2004.

Farnam, Charles H. (1846–1909). Assistant in archaeology and the first anthropology curator at Yale, 1877–91.

Ferguson, Douglas C. (1929–2002). Worked in entomology, specializing in Lepidoptera, 1963–69.

Ford, William E. (1878–1939). Assistant and curator of the Brush Mineral Collection, 1900–1939, and the mineralogy collection, 1922–39.

Fox, Marilyn. Senior preparator in vertebrate paleontology, 1997–present.

Furth, David (1945–). Curatorial associate and museum assistant, 1974–76 and 1981–83, and collections manager of entomology, 1983–89.

Gall, Lawrence F. (1956–). Systems manager for the museum and the informatics manager in entomology, 1991–present.

Galton, Peter M. (1942–). Curatorial associate in vertebrate paleontology, 1967–70.

Gauthier, Jacques A. (1951–). Curator of vertebrate paleontology, vertebrate zoology, and ornithology, 1996–present. He was one of the first to use cladistics in the study of dinosaurs.

Gibb, Hugh (1860–1932). Worked on large mounts in vertebrate paleontology and was chief preparator, 1882–1932.

Gregory, Joseph T. (1914–2007). Curator of vertebrate paleontology, 1946–60. He compiled and edited the series Bibliography of Fossil Vertebrates.

Grinnell, George B. (1849–1938). Assistant in charge of the osteology collection, 1874–80. He went on to become one of the founders of the conservation movement.

Harger, Oscar (1843–87). Assistant in paleontology specializing in Crustacea, 1872–87. He helped prepare O. C. Marsh's publications and also collected for A. E. Verrill.

Hartman, Willard D. (1921–2013). Curator of invertebrate zoology, 1953–92. He was also acting director, then director of the museum, 1987–92.

Hatcher, John B. (1861–1904). Assistant in geology, 1891–92. He was also a fossil collector for O. C. Marsh, 1884–93.

Hickey, Leo J. (1940–2013). Director of the museum, 1982–87, and curator of paleobotany and botany, 1982–2013.

Hill, Andrew P. (1946–2015). Curator of anthropology, 1992–2015, and curator of vertebrate paleontology, 2014–15.

Hole, Frank (1931–). Curator of anthropology specializing in the prehistory of the Near East, 1981–2005, and served brief tenures as acting director of the museum.

Hopson, James A. (1935–). Research assistant in vertebrate paleontology and research associate in geology specializing in synapsid evolution, 1963–67.

Hu, Shusheng (1965–). Collections manager of paleobotany, 2008–present.

Hull, Pincelli (1981–). Assistant curator of invertebrate paleontology, 2015–present.

Humphrey, Philip S. (1926–2009). Curator and researcher in ornithology, 1957–62.

Hutchinson, G. Evelyn (1903–91). Scientific consultant for the Bingham Oceanographic Laboratory, 1943–60, and the biologist on the museum's North India Expedition in 1932. He won the Verrill Medal in 1981 and is considered the father of modern ecology.

Kirsch, John A. W. (1941–2007). Assistant and associate curator of mammals, 1971–79.

Kissel, Richard (1975–). Director of public programs, 2013–present.

Klise, Linda S. (1949–). Research assistant and collections manager of paleobotany, 1984–2008.

Knight, James B. (1888–1960). Researcher in invertebrate paleontology specializing in evolution of early Mollusca, 1931–34.

Knopf, Adolph (1882–1966).Curator of the Brush Mineral Collection, 1945–51.

Lazo-Wasem, Eric A. (1957–). Assistant collections manager, then collections manager of invertebrate zoology, 1983–present.

Levin, Donald A. (1939–). Curator of botany and the Herbarium, 1969–72.

Lewis, G. Edward (1908–97). Curator of vertebrate paleontology, 1939–45.

Longwell, Chester R. (1887–1975). Advisor, then curator of geology, 1925–46. He directed the installation of a seismograph at the museum in 1926.

Lull, Richard S. (1867–1957). Curator of vertebrate paleontology, 1906–27, and acting curator of geology, 1921. He was director, then acting director of the museum, 1922–38, overseeing the construction of the current museum building and putting an evolutionary orientation into the displays.

MacClintock, Copeland (1930–). Researcher, curatorial associate, and assistant in invertebrate paleontology, 1963–2010, and assistant to the director for general administration, 1968–87.

MacCurdy, George G. (1863–1947). Curator and researcher in anthropology specializing in European prehistory, 1902–31. He conducted a survey of Native American sites across Connecticut.

Marsh, Othniel C. (1831–99). Curator of the geology collections and trustee of the museum, 1866–99. He was one of the three original curators of the museum.

McAlester, A. Lee, Jr. (1933–). Curator of invertebrate paleontology, 1959–74.

McIntosh, John S. (1923–). Research affiliate specializing in O. C. Marsh's dinosaurs and sauropods, 195? –?

McIntosh, Roderick J. (1951–). Curator of anthropology, 2007–present.

McKaye, Kenneth R. (1947–). Assistant curator of vertebrate zoology, 1975–81.

Merriman, Daniel (1908–84). Curator and director of the Bingham Oceanographic Laboratory, then director of the Sears Foundation for Marine Research, 1942–77.

Meyer, Grant E. (1935–2004). Preparator and chief preparator, 1963–77.

Miller, Arthur K. (1902–13). Researcher in invertebrate paleontology, 1929–31.

Monteiro, Antónia (1969–). Assistant curator in entomology, 2006–13. She funded and coordinated the digitization of the nymphalid butterflies at the Peabody.

Morrill, Ralph C. (1902–96). Chief preparator of the museum, 1924–66.

Morrow, James E., Jr. (1918–2002). Researcher at the Bingham Oceanographic Laboratory, 1949–60. He participated in research expeditions to New Zealand, Kenya, Peru, the Maldives, and the Seychelles.

Munstermann, Leonard E. (1942–). Curator of entomology, 1994–present.

Musto, David F. (1936–2010). Curator of historical scientific instruments, 1987–2010, and acting director of the museum, June–July 1995.

Near, Thomas J. (1969–). Associate curator of ichthyology, 2006–present.

Nicolescu, Stefan (1957–). Collections manager of mineralogy and meteoritics, 2010–present.

Norris, Christopher (1966–). Collections manager of vertebrate paleontology, 2009–present.

Osgood, Cornelius (1905–85). Curator of anthropology specializing in cultural anthropology, 1930–74, and associate director of the museum, 1966–74.

Ostrom, John H. (1928–2005). Curator of vertebrate paleontology who revolutionized the modern understanding of dinosaurs, 1961–92. He was also acting director of the museum, 1975–76.

Parr, Albert E. (1890–1991). Assistant curator of zoology, then curator and researcher at the Bingham Oceanographic Laboratory, 1927–42. He was also director of the museum, 1938–42.

Patton, William H. (1853–1918). Assistant in zoology in charge of the entomology collections, 1876–78. He was also a student and assistant of A. E. Verrill.

Penfield, Samuel L. (1856–1906). Curator of the Brush Mineral Collection, 1904–6.

Petrunkevitch, Alexander I. (1875–1964). Prominent arachnologist who received the museum's first Verrill Medal in 1959.

Pickering, Jane (1965–). Director of public programs, 2003–13, and deputy director of the museum, 2005–13.

Pickford, Grace E. (1902–86). Researcher at the Bingham Oceanographic Laboratory, 1933–55 and 1959–66, specializing in Cephalopoda, oligochaete worms, fish endocrinology, and water beetles. Yale belatedly made her full professor in 1969, the year before she retired.

Piel, William H. (1968–). Associate director of evolutionary informatics, director of informatics, and manager of the frozen tissue collection, 2006–13.

Pilbeam, David R. (1940–). Curator of anthropology, 1969–81.

Pirsson, Louis V. (1860–1919). In charge of the petrology collection, 1895–97, 1901, 1904, and appointed acting curator of geology in 1904 after C. E. Beecher's death.

Pospisil, Leopold J. (1923–). Curator of anthropology, 1956–93.

Price, Derek J. deSolla (1922–83). Curator of historical scientific instruments, 1960–83.

Prum, Richard O. (1961–). Curator of vertebrate zoology specializing in ornithology and bird evolution, 2004–present. He was awarded a MacArthur Fellowship in 2010.

Pupedis, Raymond J. (1948–). Collections manager of entomology, 1990–present.

Rainey, Froelich G. (1907–92). Assistant curator of anthropology, 1935–37, doing fieldwork in Alaska and specializing in Eskimo and Caribbean prehistory.

Ramus, Joseph S. (1940–). Assistant and associate curator of botany, 1968–78.

Reed, Charles A. (1937–92). Associate curator of mammals and reptiles specializing in the origins of animal domestication in the Old World, 1961–66.

Reeder, Charlotte G. (1916–2009). Helped curate the Herbarium while a researcher in the botany and biology departments of Yale, 1957–68.

Reeder, John R. (1914–2009). Researcher in botany, 1963–68, and a curator of the Herbarium specializing in Gramineae, 1947–68.

Remington, Charles L. (1922–2007). Researcher, then curator of entomology, 1953–92. He vastly increased the size of the entomology collection.

Rice, Sean H. (1961–). Assistant curator of invertebrate paleontology, 1995–2004.

Richard, Alison F. (1948–). Curator of anthropology, 1990–2003, acting curator of entomology, 1993, and director of the museum, 1990–94.

Richards, Sarah W. (1925–2011). Researcher at the Bingham Oceanographic Laboratory, 1949–64, research affiliate in zoology, 1972–77, and curatorial affiliate, 1977–81.

Riley, Gordon A. (1911–85). Researcher, associate director, and acting director of the Bingham Oceanographic Laboratory, 1938–63.

Ripley, S. Dillon, II (1913–2001). Curator of vertebrate zoology, then research affiliate in ornithology, 1946–75, and director and trustee of the museum, 1959–64, later becoming secretary of the Smithsonian Institution. He was awarded the Verrill Medal in 1984.

Rodman, James E. (1945–). Assistant curator of botany in charge of the Herbarium, 1973–83.

Rojas, Lourdes M. (1970–). Assistant in invertebrate zoology, 2002–present.

Rouse, B. Irving (1913–2006). Researcher, faculty affiliate, and curator of anthropology, 1938–77. He is considered the father of Caribbean archaeology.

Sapir, Edward (1884–1939). Honorary curator of anthropology specializing in linguistics and anthropology of northwestern U.S. Indians, 1931–34.

Sargis, Eric (1970–). Curator in vertebrate zoology and vertebrate paleontology, 2006–present.

Schindel, David E. (1951–). Curator of invertebrate paleontology, 1978–86.

Schuchert, Charles (1858–1942). Curator of geology and of invertebrate paleontology, 1904–26, and trustee of the museum, 1912–26. He more than doubled the invertebrate paleontology collections of the museum.

Schweitzer, Dale F. (1950–). Curatorial associate in entomology, 1975–81.

Sease, Catherine (1947–). Museum collections conservator, 2000–present.

Seilacher, Adolf (1925–2014). Adjunct curator of invertebrate paleontology, 1987–2009. He transformed our understanding of ancient organisms and showed how the burrowing of snails or other trace fossils can reveal behaviors of extinct species.

Sibley, Charles G. (1917–98). Curator of vertebrate zoology, 1965–86, and director of the museum, 1970–76. He was a pioneer of DNA-based species classification.

Sill, William D. (1937–2008). Curatorial associate in vertebrate paleontology, 1968–69.

Silliman, Benjamin (1779–1864). Established and curated Yale's earliest collections of meteorites, minerals, fossils, and other specimens. He was also a member of the Yale faculty, teaching chemistry, geology, mineralogy, and pharmacy.

Silliman, Benjamin, Jr. (1816–85). Trustee of the museum, 1866–85.

Simons, Elwyn L. (1930–). Curator of vertebrate paleontology, 1960–77.

Simpson, W. Kelly (1928–). Research associate in anthropology, 1960–?

Skelly, David (1965–). Curator of vertebrate zoology specializing in amphibians, 2000–present, and director of the museum, 2014–present.

Smith, Sidney I. (1843–1926). Assistant in zoology, 1867–74, and a member of the Yale faculty in comparative anatomy, 1874–1906. He also assisted his brother-in-law A. E. Verrill on dredging expeditions, 1864–70, and helped him set up zoological exhibits in the museum.

Stickney, Eleanor H. (1919–2011). Collection manager and librarian in ornithology, 1963–90. She was involved in the 1962 exhibition of the treasures of Tutankhamun's tomb at the museum. She was also the first president of the Peabody Museum Associates and established the volunteer program at the museum.

Sturtevant, William C. (1926–2007). Assistant curator of anthropology specializing in North American ethnology, 1954–56.

Sweeney, Patrick (1972–). Collections manager of botany, 2008–present.

Switzer, George S. (1915–2008). Assistant curator of the Brush Mineral Collection, 1940–45.

Thompson, Ernest F. (1907–95). Researcher and curator at the Bingham Oceanographic Laboratory, 1944–66.

Thomson, Keith S. (1938–). Curator of vertebrate zoology, 1965–86, who provided new insights into the biology of coelacanths. He was also a research associate at the Bingham Oceanographic Laboratory, 1965–66, acting director of the museum, 1976–77, and director of the museum, 1977–79.

Thorpe, Malcolm R. (1891–1958). Researcher and curator of vertebrate paleontology, 1919–38, and acting director of the museum, 1927.

Tiffney, Bruce H. (1949–). Curator in charge of the paleobotany collection and the Herbarium, 1977–86.

Tracy, Robert J. (1944–). Curator of mineralogy and petrology specializing in igneous and metamorphic petrology, 1981–86.

Trewin, Shae (1980–). Collections manager of historical scientific instruments, 2004–10.

Troxell, Edward L. (1884–1972). Researcher in vertebrate paleontology, 1918–25.

Turekian, Karl K. (1927–2013). Consultant, researcher, and curator of meteorites, 1960–2013. He was also acting director of the museum, August 1996 and July–August 1999.

Turner, Mary Ann (1947–). Curatorial associate, collections manager, and registrar of vertebrate paleontology, 1975–2010.

Underhill, Anne P. (1954–). Assistant curator of anthropology, 1995–99, and curator of anthropology, 2010–present, specializing in the archaeology of East Asia.

Utrup, Jessica (1980–). Assistant in invertebrate paleontology, 2006–present.

Uzzell, Thomas M., Jr. (1932–). Assistant curator of herpetology, 1967–72.

Verrill, Addison E. (1839–1926). Curator of zoology, 1867–1910, and was one of the three original curators of the museum. He was also the first member of the Yale zoology faculty.

Vrba, Elizabeth S. (1942–). Curator of vertebrate paleontology and osteology specializing in evolutionary theory, 1986–2014.

Waage, Karl M. (1915–99). Curator of invertebrate paleontology, 1946–86. He was also acting director of the museum, 1979–80, and director of the museum, 1980–82.

Wangersky, Peter J. (1927–2007). Researcher at the Bingham Oceanographic Laboratory specializing in oceanography, 1961–65.

Warren, Charles H. (1876–1950). Curator of the Brush Mineral Collection, 1939–45.

Washington, Henry S. (1867–1934). Assistant in mineralogy, 1895–96, curating the museum's mineral collection.

Watkins-Colwell, Gregory J. (1969–). Museum assistant, 2001–11, specializing in ichthyology and herpetology, then collections manager of vertebrate zoology, 2011–present.

Weaver, Charles E. (1880–1958). Researcher in invertebrate paleontology, 1933–40 and 1941–46.

White, Russell D. (Tim) (1959–). Collections manager of invertebrate paleontology, 1983–2003, and director of collections and operations, 2003–present.

Wieland, George R. (1865–1953). Lecturer and researcher in paleobotany, 1906–35. He helped amass a large collection of fossil cycadeoids and discovered the *Archelon* skeleton that is now in the Great Hall.

Williston, Samuel W. (1852–1918). Assistant in paleontology and osteology, 1876–85.

Wilson, Edmund B. (1865–1939). Assistant in zoology specializing in cytology, 1878–79.

Winchell, Horace (1915–93). Assistant curator of the Brush Mineral Collection, 1945–51, and curator of the mineral collection, 1951–85.

Yoder, Ann D. (1959–). Associate curator of mammals, 2001–5.

Zyskowski, Kristof (1966–). Collections manager in vertebrate zoology, 2001–present, specializing in ornithology and mammalogy.

ACKNOWLEDGMENTS

I COULD NOT HAVE written this book without the help of Barbara Narendra, the Peabody Museum's longtime archivist. Whenever I was stumped by a question or in need of documentation, Narendra would reply by e-mail, "Working on it," and the answer would follow soon after. Daniel Brinkman was a near rival for catching my factual errors and pointing me toward better sources. Geoffrey Giller was likewise invaluable in tracking down original documents and illustrations, putting the manuscript in order, and saving me from my mistakes.

In the Peabody Museum administration, Tim White, director of collections, understood my need for editorial independence and opened any doors that needed opening. I owe thanks also to Derek E. G. Briggs and David Skelly, the museum directors, to Jean Thomson Black, my editor at Yale University Press, to my agent John Thornton for enthusiastically supporting this project, and to Tom Lovejoy and Alison Richard for their comments on the manuscript. I am grateful to Rosemary Volpe and Robert Lorenz for their work on illustrations and photographs, and to the entire collections staff and the curators for answering my frequent questions. Thanks to Yale geologist Brian Skinner for his help on the James Dwight Dana chapter; author Roger McCoy on Schuchert's continental drift; Christopher Scotese, a geologist at the University of Texas at Arlington, on paleogeography; David Weishampel of Johns Hopkins University on John Bell Hatcher; Una Farrell of the University of Kansas on trilobites; Leo Laporte of the University of California at Santa Cruz on George Gaylord Simpson; and A. W. Crompton of Harvard University on the molecular wars era.

Thanks and apologies, finally, to my wife Karen and my dog Jack (yes, in that order, though Jack was more insistent) for my inability to do much other than write during the work on this book.

(Nashville: Vanderbilt University Press, 1971), 32–33; Hanaford, *George Peabody,* 19; Augustus J. C. Hare, *The Story of My Life* (London: Ballantyne, Hanson, 1896), 2:372.

7. Hanaford, *George Peabody,* 50; Ron Chernow, *The House of Morgan: An American Banking Dynasty and the Rise of Modern Finance* (New York: Atlantic Monthly, 1990), 3, 7.
8. Robert Charles Winthrop, *Eulogy, Pronounced at the Funeral of George Peabody, at Peabody, Massachusetts, 8 February, 1870* (Boston: John Wilson & Son, 1870), 11; Parker, *George Peabody,* 32; Caleb Marsh to George Peabody, August 18, 1834, MSS 181, box 192, folder 2, George Peabody Papers, Peabody Essex Museum, Salem, Mass.; O. C. Marsh to George Peabody, May 19, 1856, MSS 181, box 192, folder 2, George Peabody Papers, Peabody Essex Museum; Schuchert and LeVene, *O. C. Marsh,* 21, 30.
9. "Unprecedented Munificence," *Times* (London), March 26, 1862, 9.
10. *Proceedings of the Trustees of the Peabody Education Fund* (Boston: John Wilson & Son, 1875), 1:3; Earle H. West, "The Peabody Education Fund and Negro Education, 1867–1880," *History of Education Quarterly* 6, no. 2 (1966): 4, 18.
11. William L. Garrison, "Mr. Peabody and the South," *Independent,* August 16, 1869, 1.
12. Winthrop, *Eulogy,* 8, 12–13.
13. Garrison, "Mr. Peabody."

Chapter 4: The Wooing of George Peabody

1. "Reception of George Peabody," *New York Times,* October 11, 1856, 2.
2. Narendra, "Benjamin Silliman," 27; Judith P. Russell to O. C. Marsh, October 28, 1856, HM 38, reel 21, frame 587, Othniel Charles Marsh Papers, Manuscripts and Archives, Yale University Library.
3. Parker, *George Peabody,* 78; Judith P. Russell to George Peabody, July 14, 1856, MSS 181, box 192, folder 2, George Peabody Papers, Peabody Essex Museum.
4. Schuchert and LeVene, *O. C. Marsh,* 39; Hanaford, *George Peabody,* 70; Marie C. Menefee to Charles Schuchert, July 12, 1937, Vertebrate Paleontology Archives, Yale University Peabody Museum of Natural History.
5. George Peabody to O. C. Marsh, September 28, 1860, HM 38, reel 22, frame 457, Othniel Charles Marsh Papers, Manuscripts and Archives, Yale University Library.
6. Louis Agassiz, remarks on Marsh's fossil vertebrae, *American Journal of Science and Arts* 33 (1862): 138; O. C. Marsh, "Description of the Remains of a new Enaliosaurian (*Eosaurus Acadianus*), from the Coal Formation of Nova Scotia," *American Journal of Science and Arts* 32 (1862): 2.
7. O. C. Marsh to George Peabody, June 9, 1862, MSS 181, box 192, folder 2, George Peabody Papers, Peabody Essex Museum.
8. O. C. Marsh to Caleb Marsh, August 13, 1862, HM 38, reel 20, frames 737–38, Othniel Charles Marsh Papers, Manuscripts and Archives, Yale University Library.
9. Moncure D. Conway, *Autobiography: Memories and Experiences of Moncure Daniel Conway* (Boston: Houghton Mifflin, 1904), 2:281.
10. O. C. Marsh to George Peabody, October 21, 1864, HM 38, reel 22, frame 506, Othniel Charles Marsh Papers, Manuscripts and Archives, Yale University Library; Schuchert and LeVene, *O. C. Marsh,* 21; copy of letter from O. C. Marsh to Benjamin Silliman Jr., November 26, 1862, HM 38, reel 15, frame 115, Othniel Charles Marsh Papers, Manuscripts and Archives, Yale University Library. Much of Marsh's correspondence has also been digitized and is available online at http://peabody.yale.edu/collections/vertebrate-paleontology/correspondence-o-c-marsh.
11. Marsh to Silliman Jr., November 26, 1862; Benjamin Silliman Jr. to O. C. Marsh, January 20, 1863, HM 38, reel 15, frames 118–23, Othniel Charles Marsh Papers, Manuscripts and Archives, Yale University Library.
12. O. C. Marsh to Benjamin Silliman, May 25, 1863, HM 38, reel 15, frame 146; O. C. Marsh to Benjamin Silliman Jr., May 28, 1863, HM 38, reel 15, frames 151–53, Othniel Charles Marsh Papers, Manuscripts and Archives, Yale University Library.

13. James D. Dana to O. C. Marsh, June 16, 1863, MS 343, box 8, folder 321, Othniel Charles Marsh Papers, Manuscripts and Archives, Yale University Library.
14. Benjamin Silliman Jr. to O. C. Marsh, June 15, 1863, HM 38, reel 15, frames 160–67, Othniel Charles Marsh Papers, Manuscripts and Archives, Yale University Library.
15. O. C. Marsh to George Peabody, July 12, 1863, HM 38, reel 22, frames 470–74, Othniel Charles Marsh Papers, Manuscripts and Archives, Yale University Library.
16. O. C. Marsh to George J. Brush, May 5, 1864, MS 108, box 3, folder 85, George Jarvis Brush Family Papers, Manuscripts and Archives, Yale University Library; Marsh to Peabody, July 12, 1863; Schuchert and LeVene, *O. C. Marsh,* 63.
17. George J. Brush to O. C. Marsh, February 11, 1864, MS 343, box 5, folder 177; James D. Dana to O. C. Marsh, January 26, 1864, MS 343, box 8, folder 321, Othniel Charles Marsh Papers, Manuscripts and Archives, Yale University Library; Schuchert and LeVene, *O. C. Marsh,* 82–83.
18. Robert J. Richards, *The Tragic Sense of Life: Ernst Haeckel and the Struggle over Evolutionary Thought* (Chicago: University of Chicago Press, 2008), 68.
19. J. Vernon Jensen, "The X Club: Fraternity of Victorian Scientists," *British Journal for the History of Science* 5 (1970): 70.
20. Marsh, "Introduction and Succession of Vertebrate Life," 337; O. C. Marsh, *History and Methods of Palæontological Discovery: An Address Delivered Before the American Association for the Advancement of Science, at Saratoga, N. Y., August 28, 1879* (New Haven, Conn.: Tuttle, Morehouse, & Taylor, 1879), 47.
21. "Introductory Remarks of Hon. Robert C. Winthrop, Chairman of the Board of Trustees," in *Report of the Peabody Museum of American Archæology and Ethnology in Connection with Harvard University* (Salem, Mass.: Salem Press, 1878), 2:178.
22. "Mr. George Peabody's Recent Gifts to Science," *American Journal of Science and Arts* 43 (1867): 131.
23. "Mr. Peabody's Letter," October 22, 1866, RU 471, accession 19ND-A-146, box 1, folder 1, Peabody Museum of Natural History, Yale University, records, Manuscripts and Archives, Yale University Library.
24. Draft of letter from O. C. Marsh to James D. Dana, June 9, 1864, HM 38, reel 4, frames 137–44, Manuscripts and Archives, Yale University Library.
25. Parker, *George Peabody,* 147–48; *Queen Victoria: Story of Her Life and Reign 1819–1901,* Project Gutenberg EBook, 2003, http://www.gutenberg.org/files/9947/9947.txt.
26. "Funeral of George Peabody at Westminster Abbey," *New York Times,* November 13, 1869, 3; "Westminster Abbey: Mr. Peabody's Funeral," *New York Times,* November 26, 1869, 2.
27. Ziegler, "The Rocky Mountains."
28. Betts, "The Yale College Expedition of 1870"; John R. Nicholson, "Local Gossip," *St. Louis Daily Globe-Democrat,* April 10, 1887, 4; Daniel Brinkman, personal communication, December 18, 2014.

Chapter 5: Rock Render

1. Daniel Coit Gilman, *The Life of James Dwight Dana* (New York: Harper & Brothers, 1899), 170; Louis V. Pirsson, "Biographical Memoir of James Dwight Dana, 1813–1895," *National Academy of Sciences of the United States of America Biographical Memoirs* 9 (1919): 42; "Professor Dana's Excursions about New Haven," *Yale College Courant,* March 20, 1869, 167–68.
2. Pirsson, *James Dwight Dana,* 62; James Dwight Dana, "Preface to Third Edition," in *The System of Mineralogy of James Dwight Dana, 1837–1868: Descriptive Mineralogy,* 6th ed. (New York: John Wiley & Sons, 1911), viii.
3. James D. Dana to Augustus S. Gould, July 30, 1851, box 275, misc. reel 5075, frame 49, Brock Miscellaneous Files, 1655–1908, Library of Virginia, Richmond; Gilman, *James Dwight Dana,* 8.
4. Charles R. Darwin to James D. Dana, September 27, 1853, Darwin Correspondence Project, http://www.darwinproject.ac.uk/letter/entry-1533.

5. Gilman, *James Dwight Dana,* 98.
6. Ibid., 99.
7. Ibid., 99–101.
8. Ibid., 102.
9. Ibid., 101, 103.
10. Ibid., 151.
11. James Dwight Dana, *United States Exploring Expedition During the Years 1838, 1839, 1840, 1841, 1842, under the Command of Charles Wilkes, U.S.N.,* vol. 10, *Geology* (Philadelphia: C. Sherman, 1849), 105; Gilman, *James Dwight Dana,* 151, 209.
12. Gilman, *James Dwight Dana,* 118–19.
13. Charles R. Darwin to Charles Lyell, December 4, 1849, Darwin Correspondence Project, http://www.darwinproject.ac.uk/entry-1275.
14. Gilman, *James Dwight Dana,* 241, 125.
15. Dana, *United States Exploring Expedition,* 388–89.
16. David Igler, "On Coral Reefs, Volcanoes, Gods, and Patriotic Geology; or, James Dwight Dana Assembles the Pacific Basin," *Pacific Historical Review* 79 (2010): 33–34, 38.
17. James Dwight Dana, "Origin of the Grand Outline Features of the Earth," *American Journal of Science and Arts* 3 (1847): 398; Dana, *United States Exploring Expedition,* 426.
18. Michael L. Prendergast, "James Dwight Dana: The Life and Thought of an American Scientist" (Ph.D. diss., University of California, Los Angeles, 1978), 575.
19. William Stanton, "Dana, James Dwight," in *Complete Dictionary of Scientific Biography* (Detroit: Charles Scribner's Sons, 2008), 3:549–54.
20. Gilman, *James Dwight Dana,* 254.
21. Charles R. Darwin to James D. Dana, September 29, 1856, Darwin Correspondence Project, http://www.darwinproject.ac.uk/entry-1964; James D. Dana to Charles R. Darwin, December 8, 1856, Darwin Correspondence Project, http://www.darwinproject.ac.uk/letter/entry-2016.
22. Charles R. Darwin to James D. Dana, November 11, 1859, Darwin Correspondence Project, http://www.darwinproject.ac.uk/entry-2516.
23. Charles R. Darwin to Charles Lyell, December 29, 1859, Darwin Correspondence Project, http://www.darwinproject.ac.uk/entry-2612; Charles R. Darwin to James D. Dana, December 30, 1859, Darwin Correspondence Project, http://www.darwinproject.ac.uk/entry-2615.
24. James D. Dana to Charles R. Darwin, February 5, 1863, Darwin Correspondence Project, http://www.darwinproject.ac.uk/entry-3969.
25. James Dwight Dana, *Manual of Geology,* 2nd ed. (New York: Ivison, Blakeman, Taylor, 1875), 603–4.
26. Stanton, "Dana," 553; James Dwight Dana, *Manual of Geology,* 4th ed. (New York: American Book Company, 1895), 439, 1031, 1036; Stephen Jay Gould, *Leonardo's Mountain of Clams and the Diet of Worms* (New York: Harmony Books, 1998), 117.
27. Prendergast, "James Dwight Dana," 572; Gilman, *James Dwight Dana,* 264, 266.

Chapter 6: A Rumor of War

1. O. C. Marsh, "Observations on the Metamorphosis of Siredon into Amblystoma," *American Journal of Science* 46 (1868): 364, 374.
2. Henry F. Osborn, *Cope: Master Naturalist* (New York: Arno, 1978), 39.
3. Jane P. Davidson, *The Bone Sharp: The Life of Edward Drinker Cope* (Philadelphia: Academy of Natural Sciences of Philadelphia, 1997), 32.
4. Osborn, *Cope,* 157.
5. Ibid., 158; Edward D. Cope to O. C. Marsh, January 16, 1870, MS 343, box 7, folder 290, Othniel Charles Marsh Papers, Manuscripts and Archives, Yale University Library.
6. Charles H. Sternberg, *The Life of a Fossil Hunter* (New York: Henry Holt, 1909), 69; Timothy Dwight, *Memories of Yale Life and Men, 1845–1899* (New York: Dodd, Mead, 1903), 410.
7. Osborn, *Cope,* 136.
8. Menefee to Schuchert, July 12, 1937; Sargent, "Marsh Expedition"; Dwight, *Memories of Yale,* 411; O. C. Marsh to Charles W. Chandler, March 5, 1866, MSS 181, box 192, folder 2, George Peabody Papers, Peabody Essex Museum.
9. E.g., David R. Wallace, *The Bonehunters' Revenge: Dinosaurs, Greed, and the Greatest Scientific Feud of the Gilded Age* (New York: Houghton Mifflin, 1999), 277; Schuchert and LeVene, *O. C. Marsh,* 354; Dwight, *Memories of Yale,* 412.

10. Grinnell, comments on Marsh's Narrative Volume, frame 301.
11. Charles E. Beecher, "Othniel Charles Marsh," *American Journal of Science* 7 (1899): 405–6.
12. "Marsh Hurls Azoic Facts at Cope," *New York Herald*, January 19, 1890, 11.
13. Edward D. Cope, "Synopsis of the Extinct Batrachia, Reptilia, and Aves of North America, Part I," *Transactions of the American Philosophical Society* 14 (1869): 54.
14. "Marsh Hurls Azoic Facts at Cope," 11.
15. Edward D. Cope to O. C. Marsh, March 21, 1870, MS 343, box 7, folder 290, Othniel Charles Marsh Papers, Manuscripts and Archives, Yale University Library.

Chapter 7: The Marsh Expeditions

1. Marsh, Narrative Volume, frames 326–28.
2. Ibid., frames 328–29; Betts, "The Yale College Expedition of 1870," 671.
3. Marsh, Narrative Volume, frames 331–34.
4. Ibid.
5. "New and Remarkable Fossils," *College Courant*, June 15, 1872, 283; O. C. Marsh, *Odontornithes, a Monograph on the Extinct Toothed Birds of North America* (Washington, D.C.: Government Printing Office, 1880), 114.
6. Osborn, *Cope*, 160, 162; Edward D. Cope, *The Vertebrata of the Cretaceous Formations of the West* (Washington, D.C.: Government Printing Office, 1875), 48.
7. E.g., letters between E. D. Cope and O. C. Marsh, January–February 1873, HM 38, reel 3, frames 884–93, Othniel Charles Marsh Papers, Manuscripts and Archives, Yale University Library; O. C. Marsh, "On the Structure of the Skull and Limbs in Mosasauroid Reptiles, with Descriptions of New Genera and Species," *American Journal of Science and Arts* 3 (1872): 448–49.
8. Osborn, *Cope*, 184–85.
9. Samuel Smith to O. C. Marsh, January 2, 1874, MS 343, box 30, folder 1279; B. D. Smith to O. C. Marsh, August 28, 1872, MS 343, box 30, folder 1268, Othniel Charles Marsh Papers, Manuscripts and Archives, Yale University Library.
10. Osborn, *Cope*, 181.
11. Charles Kingsley, *Glaucus; or, The Wonders of the Shore* (London: Macmillan, 1890), 31.
12. O. C. Marsh, "Preliminary Description of New Tertiary Mammals," part 2, *American Journal of Science and Arts* 4 (1872): 207; Edward D. Cope, "Third Account of New Vertebrata from the Bridger Eocene of Wyoming Valley," *Palæontological Bulletins*, nos. 1–18, no. 3, 2–3; Edward D. Cope, "The Wasatch and Bridger Faunæ," in *Report of the United States Geological Survey of the Territories* (Washington, D.C.: Government Printing Office, 1884), 3:224; Wallace, *Bonehunters' Revenge*, 83–84.
13. B. D. Smith to O. C. Marsh, July 5, 1872, MS 343, box 30, folder 1268, Othniel Charles Marsh Papers, Manuscripts and Archives, Yale University Library.
14. Osborn, *Cope*, 178, 191; Joseph Leidy, "On Some New Species of Fossil Mammalia from Wyoming," *Proceedings of the Academy of Natural Sciences of Philadelphia* 24 (1872): 167–69; Robert Bakker, phone interview, November 17, 2014.
15 Osborn, *Cope*, 182.
16. Untitled article, *Chicago Inter-Ocean*, July 17, 1873, 2.
17. Wallace, *Bonehunters' Revenge*, 90.
18. Lanham, *The Bone Hunters*, 123.
19. Henry W. Farnam, recollections on Marsh sent to Ernest Howe, May 5, 1931, HM 38, reel 26, frame 462, Othniel Charles Marsh Papers, Manuscripts and Archives, Yale University Library.
20. Osborn, *Cope*, 182.
21. Samuel W. Williston, "Addenda to Part I," in *The University Geological Survey of Kansas*, vol. 4, *Paleontology: Part I, Upper Cretaceous* (Topeka: J. S. Parks, 1898), 31, 44–45.
22. O. C. Marsh, "On a New Sub-class of Fossil Birds," *American Journal of Science and Arts* 5 (1873): 162.
23. Erwin H. Barbour, "The Progenitors of Birds," in *Extract from the Proceedings of the Nebraska Ornithologists' Union at Its Third Annual Meeting* (Lincoln, Neb.: State Journal, 1902), 22; Thomas H. Huxley, "The Coming of Age of *The Origin of Species*," *Nature* 22 (1880): 3.
24. Stephen J. Gould, *Bully for Brontosaurus: Reflections*

in Natural History (New York: Norton, 1991), 169; Donald R. Prothero and Robert M. Schoch, *Horns, Tusks, and Flippers: The Evolution of Hoofed Mammals* (Baltimore: Johns Hopkins University Press, 2002), 199.

25. O. C. Marsh, "Thomas Henry Huxley," *American Journal of Science* 50 (1895): 181.
26. Leonard Huxley, *Life and Letters of Thomas Henry Huxley* (New York: D. Appleton, 1900), 495; transcript of letter from Thomas H. Huxley to Clarence King, August 19, 1876, Huxley File, Internet archive, Charles Blinderman and David Joyce, Clark University, http://aleph0.clarku.edu/huxley/letters/76.html; Thomas H. Huxley to O. C. Marsh, August 17, 1876, HM 38, reel 9, frame 490, Othniel Charles Marsh Papers, Manuscripts and Archives, Yale University Library; Marsh, "Huxley," 181.
27. Gould, *Bully for Brontosaurus,* 171.
28. "The Theory of Evolution: Prof. Huxley's Final Lecture," *New York Times,* September 23, 1876; "Professor Huxley," *New York Herald,* September 23, 1876, 5.
29. Schuchert and LeVene, *O. C. Marsh,* 236–37.
30. O. C. Marsh to Thomas H. Huxley, January 12, 1877, MS 343, box 18, folder 709, Othniel Charles Marsh Papers, Manuscripts and Archives, Yale University; Marsh, "Introduction and Succession of Vertebrate Life," 3.
31. Jens L. Franzen, *The Rise of Horses,* trans. Kirsten M. Brown (Baltimore: Johns Hopkins University Press, 2010), 101; Ernst Haeckel, *Systematische Phylogenie: Entwurf eines Natürlichen Systems der Organismen auf Grund ihrer Stammesgeschichte* (Berlin: Verlag von Georg Reimer, 1895), 549; Gould, *Bully for Brontosaurus,* 175.
32. Brian Switek, "The Branching Bush of Horse Evolution," *Laelaps* (blog), September 13, 2007, https://laelaps.wordpress.com/2007/09/13/the-branching-bush-of-horse-evolution/.
33. "The 'Yale College Expedition' of This Year," *New-York Daily Tribune,* November 17, 1873, 3.
34. Charles R. Darwin to O. C. Marsh, August 31, 1880.

Chapter 8: Professor M on the Warpath

1. Schuchert and LeVene, *O. C. Marsh,* 418–19.
2. Ibid., 176.
3. Arthur Lakes, "Journal and Notes of Natural History—Geology, Saurians, etc.," June 29, [1877], SIA RU007201, box 1, folder 1, Arthur Lakes Journals, Smithsonian Institution Archives, Washington, D.C.; also see Michael F. Kohl and John S. McIntosh, *Discovering Dinosaurs in the Old West: The Field Journals of Arthur Lakes* (Washington, D.C.: Smithsonian Institution Press, 1997), 26, 24, for a published transcript of Lakes's journals.
4. Note from Samuel W. Williston to O. C. Marsh, n.d., MS 343, box 35, folder 1523; telegram from Samuel W. Williston to O. C. Marsh, April 22, 1877, MS 343, box 35, folder 1523, Othniel Charles Marsh Papers, Manuscripts and Archives, Yale University Library.
5. General Sheridan to O. C. Marsh, May 15, 1874, HM 38, reel 15, frame 56; General Ord to O. C. Marsh, October 6, 1874, HM 38, reel 13, frames 19–20, Othniel Charles Marsh Papers, Manuscripts and Archives, Yale University Library; Mark Jaffe, *The Gilded Dinosaur* (New York: Crown), 112.
6. "Treaty Between the United States of America and Different Tribes of Sioux Indians," April 29, 1868, in *Statutes at Large of the United States of America* 15 (1867–69): 636.
7. Charles Windolph, *I Fought with Custer: The Story of Sergeant Windolph, Last Survivor of the Battle of the Little Big Horn, as Told to Frazier and Robert Hunt* (Lincoln: University of Nebraska Press, 1947), 33; George A. Custer, in "Letter from the Secretary of War," Senate Executive Documents, 43rd Cong., 2nd sess., document 32; John D. Bergamini, *The Hundredth Year: The United States in 1876* (New York: Putnam, 1976), 47; Jaffe, *Gilded Dinosaur,* 119.
8. George E. Hyde, *Red Cloud's Folk: A History of the Oglala Sioux Indians* (Norman: University of Oklahoma Press, 1975), 220–22.
9. "A Perilous Fossil Hunt," *New York Tribune* Extra, Lecture and Letter Series, no. 27, March 1875, 48; "Scientific Intelligence: 6. Return of Professor

Marsh's Expedition," *American Journal of Science and Arts* 9 (1875): 62.

10. *Report of the Special Commission Appointed to Investigate the Affairs of the Red Cloud Indian Agency, July, 1875; Together with the Testimony and Accompanying Documents* (Washington, D.C.: Government Printing Office, 1875), iii; Schuchert and LeVene, *O. C. Marsh,* 149.
11. *Report of the Special Commission,* 1–2.
12. Isaac H. Bromley to O. C. Marsh, August 23, 1875, HM 38, reel 27, frames 333–34, Othniel Charles Marsh Papers, Manuscripts and Archives, Yale University; "The Professor and the Secretary," *Christian Union,* July 21, 1875, 50.
13. *Report of the Special Commission,* lxxiv–lxxv; "The Red-Cloud Report," *Nation,* October 28, 1875, 273; "The Interior Department," *Boston Evening Transcript,* October 19, 1875, 4.
14. William L. Carpenter to O. C. Marsh, January 31, 1877, MS 343, box 6, folder 216, Othniel Charles Marsh Papers, Manuscripts and Archives, Yale University Library.

Chapter 9: The Year of Enormous Dinosaurs

1. Lakes, "Journal and Notes of Natural History"; Kohl and McIntosh, *Discovering Dinosaurs,* 10.
2. Lakes, "Journal and Notes of Natural History"; Kohl and McIntosh, *Discovering Dinosaurs,* 10–12.
3. Arthur Lakes to O. C. Marsh, April 2, 1877, and April 20, 1877, MS 343, box 20, folder 799; telegram from Benjamin F. Mudge to O. C. Marsh, June 30, 1877, MS 343, box 24, folder 968, Othniel Charles Marsh Papers, Manuscripts and Archives, Yale University Library.
4. O. C. Marsh, "Notice of a New and Gigantic Dinosaur," *American Journal of Science* 14 (1877): 87–88; O. C. Marsh, "A New Order of Extinct Reptilia (Stegosauria) from the Jurassic of the Rocky Mountains," *American Journal of Science* 14 (1877): 513.
5. Lakes, "Journal and Notes of Natural History"; Kohl and McIntosh, *Discovering Dinosaurs,* 68–69.
6. "Harlow and Edwards" (William Harlow Reed and William Edwards Carlin) to O. C. Marsh, July 19, 1877, MS 343, box 27, folder 1125; Samuel W. Williston to O. C. Marsh, November 14, 1877, MS 343, box 35, folder 1525, Othniel Charles Marsh Papers, Manuscripts and Archives, Yale University Library.
7. Samuel W. Williston to O. C. Marsh, November 14, 1877, MS 343, box 35, folder 1525; William H. Reed to Samuel W. Williston, March 22, 1878, MS 343, box 27, folder 1126, Othniel Charles Marsh Papers, Manuscripts and Archives, Yale University Library.
8. John H. Ostrom and John S. McIntosh, *Marsh's Dinosaurs: The Collections from Como Bluff* (New Haven, Conn.: Yale University Press, 1999), xv.
9. Edward D. Cope, "On a Gigantic Saurian from the Dakota Epoch of Colorado," *Paleontological Bulletin,* no. 25, August 23, 1877, 10.
10. D. Cary Woodruff and John R. Foster, "The Fragile Legacy of *Amphicoelias fragillimus* (Dinosauria: Sauropoda; Morrison Formation—Latest Jurassic)," *Volumina Jurassica* 12 (2014): 218, 212.
11. Samuel W. Williston to O. C. Marsh, December 9, 1877, MS 343, box 35, folder 1525; Samuel W. Williston to O. C. Marsh, September 26, 1877, MS 343, box 35, folder 1524, Othniel Charles Marsh Papers, Manuscripts and Archives, Yale University Library.
12. Wallace, *Bonehunters' Revenge,* 115; W. Barksdale Maynard, "Tigers and Dinosaurs," *Princeton Alumni Weekly,* November 13, 2013, https://paw.princeton.edu/issues/2013/11/13/pages/6409/index.xml?page=2&; "Scientists Wage Bitter Warfare," *New York Herald,* January 12, 1890, 10; William H. Reed to Samuel W. Williston, February 12, 1879, MS 343, box 17, folder 1127, Othniel Charles Marsh Papers, Manuscripts and Archives, Yale University Library.
13. Wallace, *Bonehunters' Revenge,* 113; "Professor Marsh's Address," *Yale Alumni Weekly,* February 24, 1898: 7; D. Jerome Fisher, *The Seventy Years of the Department of Geology, University of Chicago, 1892–1961* (Chicago: University of Chicago, 1963), 18.
14. "Gigantic Reptiles and Dragons," *New York Times,* July 17, 1878, 3 (reprinted from the *Omaha [Neb.] Bee*).

15. Russell H. Chittenden, in Charles Schuchert, "Biographical Memoir of Othniel Charles Marsh, 1831–1899," *National Academy of Sciences of the United States of America Biographical Memoirs* 20 (1938): 28–29; Ernest Howe, "O. C. Marsh" [likely comments prepared for 1931 address], HM 38, reel 26, frame 353, Othniel Charles Marsh Papers, Manuscripts and Archives, Yale University Library.
16. Schuchert and LeVene, "O. C. Marsh," 308.
17. Elizabeth N. Shor, *Fossils and Flies: The Life of a Compleat Scientist, Samuel Wendell Williston (1851–1918)* (Norman: University of Oklahoma Press, 1971), 101.
18. William B. Scott, *Some Memories of a Palaeontologist* (Princeton, N.J.: Princeton University Press, 1939), 58; Wallace E. Stegner, *Beyond the Hundredth Meridian: John Wesley Powell and the Second Opening of the West* (Boston: Houghton Mifflin, 1954), 284.
19. Osborn, *Cope,* 585; Henry F. Osborn, *Impressions of Great Naturalists: Reminiscences of Darwin, Huxley, Balfour, Cope and Others* (New York: Charles Scribner's Sons, 1924), 161–62.
20. Osborn, *Cope,* 585.

SIDEBAR: *BRONTOSAURUS* REDUX

a. Arthur Lakes, July 25, 1879, entry, "Journal of Saurian Hunting in Wyoming, 1879," SIA RU007201, box 1, folder 3, Arthur Lakes Journals, Smithsonian Institution Archives, Washington, D.C.
b. Elmer S. Riggs, *Structure and Relationships of Opisthocœlian Dinosaurs, Part I: Apatosaurus Marsh,* Field Columbian Museum publication 82 (Chicago, 1903), 196.
c. Gould, *Bully for Brontosaurus,* 92.
d. Jacques Gauthier, speech, Yale Peabody Museum of Natural History, New Haven, Conn., April 14, 2015.
e. Emanuel Tschopp, Octávio Mateus, and Roger B. J. Benson, "A Specimen-Level Phylogenetic Analysis and Taxonomic Revision of Diplodocidae (Dinosauria, Sauropoda)," *PeerJ,* April 7, 2015, doi: 10.7717/peerj.857.
f. Richard Conniff, "Scientists Rediscover Disney's Dinosaur," *TakePart,* April 7, 2015, http://www.takepart.com/article/2015/04/07/scientists-rediscover-brontosaurus-dinosaur; Gauthier, speech.

Chapter 10: Fossils, Buffalo, and the Birth of American Conservation

1. Richard Vaughan, "Broad Are Nebraska's Rolling Plains: The Early Writings of George Bird Grinnell," *Faculty Publications: Indiana University Law Library,* paper 714 (2002), 42; Grinnell, comments on Marsh's Narrative Volume.
2. George Bird Grinnell, "Some Audubon Letters," *Auk* 33 (1916): 119–20.
3. George Bird Grinnell, *The Passing of the Great West: Selected Papers of George Bird Grinnell,* ed. John F. Reiger (New York: Winchester, 1972), 27.
4. George Bird Grinnell, "Memoirs," HM 223, reel 46, frame 276, George Bird Grinnell Papers, Manuscripts and Archives, Yale University Library.
5. Grinnell, *The Passing of the Great West,* 55–56.
6. George Bird Grinnell (writing as "Ornis"), "Buffalo Hunt with the Pawnees," *Forest and Stream,* December 25, 1873, 305; Michael Punke, *Last Stand: George Bird Grinnell, the Battle to Save the Buffalo, and the Birth of the New West* (Lincoln: University of Nebraska Press, 2009), 39; George Bird Grinnell, *Pawnee Hero Stories and Folk-tales: With Notes on the Origin, Customs and Character of the Pawnee People* (New York: Charles Scribner's Sons, 1912), 296–301.
7. E.g., George Catlin, *North American Indians* (New York: Viking Penguin, 1989), 259–60; Grinnell, "Buffalo Hunt."
8. George A. Custer, dispatch of July 15, 1874, in *Pages from Black Hills Expedition Order & Dispatch Book, July 1, to August 25, 1874,* WA MSS 128, Beinecke Rare Book and Manuscript Library, Yale University; John F. Reiger, "With Grinnell and Custer in the Black Hills," *Discovery* 20, no. 1 (1987): 20; George B. Grinnell, *Two Great Scouts and Their Pawnee Battalion* (Cleveland: Arthur H. Clark, 1928), 242; George Bird Grinnell, account of 1874 Black Hills expedition, HM 223, reel 46, frame 974, George Bird Grinnell Papers, Manuscripts and Archives, Yale University Library.

9. Reiger, "Grinnell and Custer," 18–21; William Ludlow, *Report of a Reconnaissance of the Black Hills of Dakota, Made in the Summer of 1874* (Washington, D.C.: Government Printing Office, 1875), 84.
10. Grinnell, "Memoirs," frame 288; Ludlow, *Reconnaissance of the Black Hills,* 19.
11. George Bird Grinnell, "Letter of Transmittal," in William Ludlow, *Report of a Reconnaissance from Carroll, Montana Territory, on the Upper Missouri, to Yellowstone National Park, and Return, Made in the Summer of 1875* (Washington, D.C.: Government Printing Office, 1876), 61.
12. Grinnell, *The Passing of the Great West,* 125; Grinnell, account of Black Hills expedition, frame 978.
13. Grinnell, "Memoirs," frame 311.
14. George Bird Grinnell (writing as "Yo"), "A Trip to North Park," *Forest and Stream,* October 30, 1879, 771.
15. George B. Grinnell to Luther H. North, February 17, 1887, HM 223, reel 1, frame 97; George B. Grinnell to Madison Grant, January 4, 1918, HM 223, reel 23, frame 240, George Bird Grinnell Papers, Manuscripts and Archives, Yale University Library.
16. George Bird Grinnell, ed., *Hunting at High Altitudes: The Book of the Boone and Crockett Club* (New York: Harper & Brothers, 1913), 451.
17. Ibid., 451, 436; Punke, *Last Stand,* 214.
18. Punke, *Last Stand,* 217.
19. George Bird Grinnell, "The Audubon Society," *Forest and Stream,* February 11, 1886, 41; Celia L. Thaxter, "Woman's Heartlessness," *Audubon Magazine* 1 (1887–88), 13; Carolyn Merchant, "George Bird Grinnell's Audubon Society: Bridging the Gender Divide in Conservation," *Environmental History* 15 (January 2010), 21.
20. "A Plank," *Forest and Stream,* February 3, 1894, 1.
21. Garland E. Allen, "'Culling the Herd': Eugenics and the Conservation Movement in the United States, 1900–1940," *Biology Faculty Publications: Washington University,* paper 6 (2013): 53; Adam Rome, "Nature Wars, Culture Wars: Immigration and Environmental Reform in the Progressive Era," *Environmental History* 13 (2008), 434.
22. Sherry Lynn Smith, *Reimagining Indians: Native Americans Through Anglo Eyes, 1880–1940* (Cary, N.C.: Oxford University Press, 2000), 46–47.

Chapter 11: A Building of Their Own

1. James D. Dana, Benjamin Silliman Jr., George J. Brush, and O. C. Marsh to President and Fellows of Yale College, July 16, 1869, Records of the Trustees of the Peabody Museum, Yale Peabody Museum of Natural History Archives.
2. Hand-drawn map by James D. Dana, Records of the Trustees, Yale Peabody Museum of Natural History Archives.
3. "The Peabody Museum, Yale College," *Frank Leslie's Illustrated Newspaper,* January 26, 1878, 361.
4. "The Peabody Museum, Yale College, New Haven, Conn." *Scientific American,* supplement 3, September 8, 1877, 1400.
5. Ibid., 1400–1401.
6. Schuchert and LeVene, *O. C. Marsh,* 91, 296; Schuchert, "Biographical Memoir of Othniel Charles Marsh, 1831–1899," 1.
7. "The Peabody Museum, Yale College," in *The Western Review of Science and Industry,* Vol. I, ed. Theo S. Case (Kansas City, Mo.: Journal of Commerce, 1877–88), 1:404; "The Peabody Museum," *Scientific American,* 1401.
8. Karl Alfred von Zittel, "Museums of Natural History in the United States," *Science,* February 15, 1884, 194 (translated from the supplement to the December 16, 1883, issue of *Allgemeine Zeitung*).
9. Ibid.
10. Ibid.
11. Ibid.
12. Marsh to Dana, June 9, 1864; P. T. Barnum to O. C. Marsh, February 3, 1886, MS 343, box 2, folder 61, Othniel Charles Marsh Papers, Manuscripts and Archives, Yale University Library.
13. Kathleen A. Curran, "A Forgotten Architect of the Gilded Age: Josiah Cleaveland Cady's Legacy," *Watkinson Exhibition Catalogs,* paper 20 (1993), 13, http://digitalrepository.trincoll.edu/exhibitions/20.

SIDEBAR: MR. MARSH BUILDS HIS DREAM HOUSE

a. Schuchert and LeVene, *O. C. Marsh,* 348.

Chapter 12: In the Shadow of O. C. Marsh

1. Moses M. Harvey, "How I Discovered the Great Devil-Fish," *Wide World Magazine* 2 (1898–99): 732.
2. Jules Verne, *Twenty Thousand Leagues under the Sea* (New York: Butler Brothers, 1887), 314; Victor Hugo, *Toilers of the Sea* (New York: Harper & Brothers, 1867), 123; Addison E. Verrill, *Report on the Cephalopods of the Northeastern Coast of America* (Washington, D.C.: Government Printing Office, 1882), 6; Harvey, "Great Devil-Fish," 732–33.
3. Harvey, "Great Devil-Fish," 739.
4. Wesley R. Coe, "Biographical Memoir of Addison Emery Verrill, 1839–1926," *National Academy of Sciences of the United States of America Biographical Memoirs* 14 (1929): 27, 39.
5. Christoph Irmscher, *Louis Agassiz: Creator of American Science* (New York: Houghton Mifflin, 2013), 195; diary of Addison E. Verrill, January 1, 1860–December 21, 1863, HUD 860.90, Harvard University Archives; Samuel H. Scudder, "In the Laboratory with Agassiz," *Every Saturday,* April 4, 1874, 370; Lane Cooper, *Louis Agassiz as a Teacher: Illustrative Extracts on His Method of Instruction* (Ithaca, N.Y.: Comstock, 1917), 1; diary of Verrill, February 24–27, 1863; Coe, "Addison Emery Verrill," 26.
6. Diary of Verrill, September 11, 1860, and September 29, 1860; Irmscher, *Louis Agassiz,* 130.
7. Diary of Verrill, December 12, 1862; Richard I. Johnson, "Molluscan Taxa of Addison Emery Verrill and Katharine Jeannette Bush, Including Those Introduced by Sanderson Smith and Alpheus Hyatt Verrill," *Occasional Papers on Mollusks,* August 30, 1989, 4.
8. Addison E. Verrill to O. C. Marsh, October 22, 1871, box 33, folder 1413, Othniel Charles Marsh Papers, Manuscripts and Archives, Yale University Library.
9. G. Brown Goode, "The First Decade of the United States Fish Commission: Its Plan of Work and Accomplished Results, Scientific and Economical," *Bulletin of the U. S. Fish Commission* (Washington, D.C.: Government Printing Office, 1883), 2:170.
10. Addison E. Verrill and Sidney I. Smith, *Report upon the Invertebrate Animals of Vineyard Sound and Adjacent Waters, with an Account of the Physical Features of the Region* (Washington, D.C.: Government Printing Office, 1874), 95–96; Paul S. Galtsoff, *The Story of the Bureau of Commercial Fisheries Biological Laboratory, Woods Hole, Massachusetts,* circular 145 (Washington, D.C.: U.S. Department of the Interior, May 25, 1962), 23.
11. Joel W. Hedgpeth, introduction to "Treatise on Marine Ecology and Paleoecology, Volume 1: Ecology," *Geological Society of America Memoirs* 67V1 (1957), 4–5; Verrill and Smith, *Animals of Vineyard Sound,* 5.
12. "Life in the Gulf Stream," *New York Times,* October 29, 1882; A. D. Mead, "The Natural History of the Star-fish," in *Bulletin of the United States Fish Commission,* vol. 19 (Washington, D.C.: Government Printing Office, 1901), 224.
13. Addison E. Verrill, "A Very Valuable and Unique Zoological Collection for Sale" (pamphlet), IZAR.001777, Addison Emery Verrill Archives, Invertebrate Zoology Division, Yale Peabody Museum of Natural History Archives.
14. "Yale Buys Collection," *New York Times,* May 17, 1908; George E. Verrill, *The Ancestry, Life and Work of Addison E. Verrill of Yale University* (Santa Barbara, Calif.: Pacific Coast, 1958), 58.
15. Minutes of the Peabody Trustees, February 28, 1907, Records of the Trustees, vol. 1 (1866–1917), 147–48, Yale Peabody Museum of Natural History Archives; copy of letter from Edward S. Dana to Addison E. Verrill, May 18, 1909, Records of the Trustees, vol. 1, 156.
16. Addison E. Verrill to Ross G. Harrison, April 16, 1925, MS 263, box 33, folder 302, Ross Granville Harrison Papers, Manuscripts and Archives, Yale University Library.
17. G. E. Verrill, *Life of A. E. Verrill,* 76; William H. Dall to Ellen M. Wilde, August 1, 1866, SIA RU007073, box 4, folder 1, Letters, 1866–1871, p. 70,

William H. Dall Papers, Smithsonian Institution Archives.

18. G. E. Verrill, *Life of A. E. Verrill,* 76.

SIDEBAR: A VISIBLE WOMAN

a. Bruce Weber, "Conchita Cintrón, 'Goddess' of Bullring, Dies at 86," *New York Times,* February 21, 2009.

b. Coe, "Biographical Memoir of Verrill," 39.

c. Jeanne Remington, "Katharine Jeannette Bush: Peabody's Mysterious Zoologist," *Discovery* 12 (1977): 3; Katharine J. Bush, "Notes on the Family Pyramidellidae," *American Journal of Science* 27 (1909): 475–84; Ruth D. Turner to Jeanne E. Remington, March 1, 1978, Katharine Jeannette Bush biographical file, Yale Peabody Museum of Natural History Archives, New Haven, Conn.

d. Mrs. John A. Logan, *The Part Taken by Women in American History* (Wilmington, Del.: Perry-Nalle, 1912), 880.

Chapter 13: The Prince of Bone Hunters

1. Lowell Dingus and Mark A. Norell, *Barnum Brown: The Man Who Discovered* Tyrannosaurus rex (Berkeley: University of California Press, 2010), xiii.
2. Charles A. Guernsey, *Wyoming Cowboy Days* (New York: G. P. Putnam's Sons, 1936), 113–14.
3. Guernsey, *Wyoming Cowboy Days,* 114–15; Schuchert and LeVene, *O. C. Marsh,* 212–15; John B. Hatcher, "The Ceratopsia," in *Monographs of the United States Geological Survey,* vol. 49 (Washington, D.C.: Government Printing Office, 1907), 8.
4. John B. Hatcher to O. C. Marsh, May 7, 1889, MS 343, box 15, folder 613, Othniel Charles Marsh Papers, Manuscripts and Archives, Yale University Library.
5. John B. Hatcher to O. C. Marsh, July 10, 1884, MS 343, box 15, folder 602, Othniel Charles Marsh Papers, Manuscripts and Archives, Yale University Library.
6. John B. Hatcher to O. C. Marsh, August 17, 1884, MS 343, box 15, folder 602, Othniel Charles Marsh Papers, Manuscripts and Archives, Yale University Library.
7. William J. Holland, "John Bell Hatcher," *Geological Magazine* 1 (1994): 571; John B. Hatcher, "The Ceratops Beds of Converse County, Wyoming," *American Journal of Science* 45 (1893): 135, 143.
8. William B. Scott, "John Bell Hatcher," *Science,* July 29, 1904, 139; John B. Hatcher to O. C. Marsh, December 14, 1891, MS 343, box 16, folder 624; John B. Hatcher to O. C. Marsh, June 27, 1889, MS 343, box 15, folder 613, Othniel Charles Marsh Papers, Manuscripts and Archives, Yale University Library; Henry F. Osborn, "Explorations of John Bell Hatcher," in Hatcher, "The Ceratopsia."
9. Charles E. Beecher to O. C. Marsh, July 20, 1889, MS 343, box 3, folder 89, Othniel Charles Marsh Papers, Manuscripts and Archives, Yale University Library; Holland, "Hatcher," 572.
10. Beecher to Marsh, July 20, 1889.
11. John B. Hatcher to O. C. Marsh, July 21, 1889, MS 343, box 15, folder 613, Othniel Charles Marsh Papers, Manuscripts and Archives, Yale University Library; Scott, *Some Memories of a Palaeontologist,* 186; John B. Hatcher to O. C. Marsh, October 7, 1891, MS 343, box 16, folder 624, Othniel Charles Marsh Papers, Manuscripts and Archives, Yale University Library.
12. O. C. Marsh, "The Skull of the Gigantic Ceratopsidae," *American Journal of Science* 38 (1889): 501–6; John B. Hatcher to O. C. Marsh, August 17, 1889, MS 343, box 15, folder 614, Othniel Charles Marsh Papers, Manuscripts and Archives, Yale University Library.
13. John B. Hatcher to O. C. Marsh, August 31, 1890, MS 343, box 15, folder 618, Othniel Charles Marsh Papers, Manuscripts and Archives, Yale University Library.
14. John B. Hatcher, *Reports of the Princeton University Expeditions to Patagonia, 1896–1899,* vol. 1, *Narrative and Geography,* ed. William B. Scott (Princeton, N.J.: Princeton University, 1903), 14, 189.
15. Ibid., 89; Dingus and Norell, *Barnum Brown,* 65.
16. James T. Duce, "Patter-Gonia," *Atlantic Monthly,* September 1937, 372.
17. Hatcher, *Reports of the Princeton University Expeditions,* dedication page.

Chapter 14: Bone Wars

1. Untitled editorial, *New York Times,* November 9, 1885, 4; Osborn, *Cope,* 380.
2. Osborn, *Cope,* 380, 381–82.
3. "Geological Survey Abuses," *New York Times,* September 16, 1885, 1.
4. Osborn, *Cope,* 304; Edward S. Dana, Charles Schuchert, et al., *A Century of Science in America, with Special Reference to the "American Journal of Science," 1818–1918* (New Haven, Conn.: Yale University Press, 1918), 237.
5. Osborn, *Cope,* 381–82.
6. Henry F. Osborn, "Biographical Memoir of Edward Drinker Cope, 1840–1897," *National Academy of Sciences of the United States of America Biographical Memoirs* 13 (1929): 135–36.
7. Osborn, *Cope,* 379; Wallace, *Bonehunters' Revenge,* 207.
8. "Scientists Wage Bitter Warfare," 10.
9. Ibid.
10. "An Old Grievance Aired," *New York Times,* January 13, 1890, 8; Elizabeth N. Shor, *The Fossil Feud Between E. D. Cope and O. C. Marsh* (Hicksville, N.Y.: Exposition, 1974), 50; Osborn, *Cope,* 410.
11. "Scientists Wage Bitter Warfare," 10.
12. Ibid.
13. Ibid.
14. Osborn, *Cope,* 403–4; "Scientists Wage Bitter Warfare," 10–11; "Volley for Volley in the Great Scientific War," *New York Herald,* January 13, 1890, 4.
15. "Scientists Wage Bitter Warfare," 11.
16. "Volley for Volley," 4.
17. "Marsh Hurls Azoic Facts at Cope," 11.
18. Ibid.
19. Ibid.
20. Ibid.
21. "War among the Scientists," *Chicago Tribune,* January 16, 1890, 4.
22. "Cope May Be Removed," *Philadelphia Inquirer,* January 14, 1890, 2; Osborn, *Cope,* 408–10.
23. Davidson, *The Bone Sharp,* 108; Osborn, *Cope,* 462–68.
24. Stegner, *Beyond the Hundredth Meridian,* 343.
25. Randy Moore, Mark Decker, and Sehoya Cotner, *Chronology of the Evolution-Creation Controversy* (Santa Barbara, Calif.: Greenwood, 2010), 133; Schuchert and LeVene, *O. C. Marsh,* 318–19.
26. Schuchert and LeVene, *O. C. Marsh,* 322–23.
27. Ronald Rainger, *An Agenda for Antiquity: Henry Fairfield Osborn & Vertebrate Paleontology at the American Museum of Natural History, 1890–1935* (Tuscaloosa: University of Alabama Press, 1991), 23, 81, 83.
28. Schuchert and LeVene, *O. C. Marsh,* 378; "Cuvier Prize Given to Yale," December 25, 1897, *New York Times,* 1.
29. "Professor Marsh's Address"; Schuchert and LeVene, *O. C. Marsh,* 325.
30. Schuchert and LeVene, 330–31.

Chapter 15: Trilobite Magic and Cycad Obsessions

1. Michael E. Taylor and Richard A. Robison, "Trilobites in Utah Folklore," *Brigham Young University Geology Studies* 23 (1976): 2, gives this translation: "little water bug like stone house in."
2. Charles D. Walcott, "Notes on some Appendages of the Trilobites," *Proceedings of the Biological Society of Washington* 9 (1894–95): 89; William S. Valiant, "Appendaged Trilobites," *Mineral Collector* 8 (1901): 111.
3. Russell H. Chittenden, "An Appreciation by the Director of the Sheffield Scientific School," *Yale Alumni Weekly,* March 2, 1904, 488; Charles E. Beecher to O. C. Marsh, July 4, 1893, MS 343, box 3, folder 89, Othniel Charles Marsh Papers, Manuscripts and Archives, Yale University Library.
4. William H. Dall, "Biographical Memoir of Charles Emerson Beecher, 1856–1904," *National Academy of Sciences of the United States of America Biographical Memoirs* 12 (1904): 63.
5. Charles Schuchert, foreword to "The Appendages, Anatomy, and Relationships of Trilobites," *Memoirs of the Connecticut Academy of Sciences* 7 (1920): 6.
6. Derek E. G. Briggs and Gregory D. Edgecombe, "The Gold Bugs," *Natural History* 11 (1992): 37.
7. George R. Wieland, "Dr. Wieland and Professor Marsh" (letter to the editor), *Yale Alumni Weekly,* March 18, 1908, 616–17.
8. Richard S. Lull, "Reminiscences of the Collec-

tion of Vertebrate Paleontology in the Peabody Museum, Yale University," 1946, Vertebrate Paleontology Archives, Yale University Peabody Museum of Natural History.

9. George R. Wieland to O. C. Marsh, September 6, 1898, and September 28, 1898, MS 343, box 35, folder 1506, Othniel Charles Marsh Papers, Manuscripts and Archives, Yale University Library.
10. George R. Wieland, *American Fossil Cycads* (Washington, D.C.: Carnegie Institution, 1906), 6.
11. Ibid., 45; minutes of meeting, April 26, 1900, Records of the Trustees, vol. 1, 98.
12. Vincent L. Santucci and John M. Ghist, "Fossil Cycad National Monument: A History from Discovery to Deauthorization," Proceedings of the 10th Conference of Fossil Resources, Rapid City, S.D., May 2014, *Dakoterra* 6 (2014): 86.
13. George R. Wieland, "Fossil Cycad National Monument," *Science* 85 (1937): 289.
14. Carl O. Dunbar, interview by Karl M. Waage, February 1–2, 1971, Dunedin, Fla., transcript, Yale University Peabody Museum of Natural History Archives.
15. Richard E. Harrison, letter to the editor, *Time,* September 6, 1937.
16. Ibid.
17. Spencer F. Baird to O. C. Marsh, November 22, 1875, MS 343, box 1, folder 41; unsigned copy of letter from O. C. Marsh to Spencer F. Baird, December 20, 1875, MS 343, box 1, folder 41, Othniel Charles Marsh Papers, Manuscripts and Archives, Yale University Library.
18. Schuchert and LeVene, *O. C. Marsh,* 383.
19. Charles E. Beecher, "Reconstruction of a Cretaceous Dinosaur, *Claosaurus annectens* Marsh," *Transactions of the Connecticut Academy of Arts and Sciences* 11 (1901–2): 312–13.
20. "A Dinosaur Restored," *New York Times,* April 23, 1901, 1; "The Solution," *Yale Alumni Weekly,* March 7, 1913, 624.
21. Rainger, *An Agenda for Antiquity,* 85.
22. Ibid., 85–87.

Chapter 16: Mapping Ancient Worlds

1. "Report of the President of Yale University," *Bulletin of Yale University,* 1st ser., no. 4 (June 1905): 11.
2. Minutes of meeting of the Executive Committee, June 8, 1905, Records of the Trustees, vol. 1, 133.
3. Charles Schuchert, copy of "Remarks of Charles Schuchert on Receiving the Mary Clark Thompson Medal at the Dinner of the Academy" (speech, Cleveland, Ohio, November 20, 1934), ms. in the possession of Russell D. White, director of collections and operations, Peabody Museum.
4. Ibid.; Karl Becker, "Cincinnati Area, the Mother of Geologists," *Compass of Sigma Gamma Epsilon* 19 (1938): 193.
5. Robert H. Dott Jr., "James Hall Jr., 1811–1898," *National Academy of Sciences of the United States of America Biographical Memoirs* 87 (2005): 8–9.
6. Adolph Knopf, "Charles Schuchert, 1858–1942," *National Academy of Sciences of the United States of America Biographical Memoirs* 27 (1952): 370.
7. Russell D. White, personal communication, May 12, 2015.
8. Knopf, "Charles Schuchert," 369–72; Philip B. King, "Atlas of Paleogeographic Maps of North America," review, *Journal of Paleontology* 30 (1956): 986.
9. Eliot Blackwelder, "Paleogeography of North America," review, *Science* 31 (1910): 910–12.
10. Charles Schuchert to Richard S. Lull, January 6, 1906, MS 435, box 31, book 2, p. 503, Charles Schuchert Papers, Manuscripts and Archives, Yale University Library; Malcolm P. Weiss and Ellis L. Yochelson, "Ozarkian and Canadian Systems: Gone and Nearly Forgotten," *Ordovician Odyssey: Short Papers for the Seventh International Symposium on the Ordovician System* (1995): 41–44.
11. Charles Schuchert, "My Presidency of the Geological Society of America," 1922, MS in the possession of Russell D. White, director of collections and operations, Peabody Museum.
12. Malcolm P. Weiss, "Geological Society of America Election of 1921: Attack on Candidacy of Charles Schuchert for the Presidency," *Earth Sciences History* 11 (1992): 94; Charles Schuchert to George P. Merrill, April 8, 1821, MS 435, box 36, book 1, pp. 898–99, Charles Schuchert Papers, Manuscripts and Archives, Yale University Library.

13. Weiss, "Attack on Candidacy of Charles Schuchert," 92–93.
14. Ibid., 96–97.
15. Ibid., 98; Malcolm P. Weiss and Russell D. White, "Geological Society of America Election of 1921: A Reprise," *Earth Sciences History* 17 (1998): 30.
16. *Report of the President of Yale University for the Year Ending December 31, 1897,* 1898, HM 214, reel 1, Manuscripts and Archives, Yale University Library; Charles Schuchert, "The Rise of Natural History Museums in the United States," in "Addresses Delivered on the Occasion of the Dedication of the New Museum Building, 29 December 1925," *Peabody Museum of Natural History Bulletin* 1 (1926): 22.
17. "Peabody Museum: Report of the Curators," *Bulletin of Yale University,* 12th ser., no. 10 (July 1916): 337.
18. Ibid.
19. Edward S. Dana to Charles Schuchert, August 30, 1917, MS 435, box 11, folder 95, Charles Schuchert Papers, Manuscripts and Archives, Yale University Library.

SIDEBAR: MISSING THE DRIFT

Parts of this sidebar are adapted from Richard Conniff, "When Continental Drift Was Considered Pseudoscience," *Smithsonian,* June 2012, 36–38.

a. Roger M. McCoy, *Ending in Ice: The Revolutionary Idea and Tragic Expedition of Alfred Wegener* (Oxford: Oxford University Press, 2006), 19.
b. Ibid., 26.
c. Anthony Hallam, *A Revolution in the Earth Sciences: From Continental Drift to Plate Tectonics* (London: Oxford University Press, 1974), 9; Charles Schuchert to Alexander du Toit, April 25, 1922, MS 435, box 37, p. 622, Charles Schuchert Papers, Manuscripts and Archives, Yale University Library, New Haven, Conn.
d. Charles Schuchert, "The Hypothesis of Continental Displacement," in *Theory of Continental Drift: A Symposium* (Tulsa: American Association of Petroleum Geologists, 1928), 139.
e. McCoy, *Ending in Ice,* 39–40.
f. Schuchert, "Continental Displacement," 140.
g. Naomi Oreskes, *The Rejection of Continental Drift: Theory and Method in American Earth Science* (New York: Oxford University Press, 1999), 178–79.
h. Charles Schuchert and Carl O. Dunbar, *A Textbook of Geology,* part 2, *Historical Geology* (New York: John Wiley & Sons, 1933), 2.

Chapter 17: A City Raised Like a Chalice

1. Jorge A. Flores Ochoa, "Contemporary Significance of Machu Picchu," trans. Richard L. Burger, in *Machu Picchu: Unveiling the Mystery of the Incas,* ed. Richard L. Burger and Lucy C. Salazar (New Haven, Conn.: Yale University Press, 2004), 109.
2. Hiram Bingham III, *Across South America: An Account of a Journey from Buenos Aires to Lima by Way of Potosí* (Boston: Houghton Mifflin, 1911), 274, 291.
3. Hiram Bingham III to Hiram Bingham II, January 10, 1907, MS 81, box 6, folder 85, Bingham Family Papers, Manuscripts and Archives, Yale University Library.
4. Alfred M. Bingham, *Portrait of an Explorer: Hiram Bingham, Discoverer of Machu Picchu* (Ames: Iowa State University Press, 1989), 118.
5. Hiram Bingham III to Alfreda M. Bingham, February 13(?), 1911, MS 81, box 15, folder 38, Bingham Family Papers.
6. Bingham, *Portrait,* 133.
7. Ibid., 97, 108, 137.
8. Hiram Bingham III, "The Discovery of Machu Picchu," *Harper's Monthly* 127 (1913): 709.
9. Bingham, *Portrait,* 156.
10. Field notebook of Hiram Bingham III, 41, MS 664, box 18, folder 1, Yale Peruvian Expedition Papers, Manuscripts and Archives, Yale University Library; Hiram Bingham III, *Inca Land: Explorations in the Highlands of Peru* (Boston: Houghton Mifflin, 1922), 315. He appears to have borrowed the phrase "capable of making considerable springs when in pursuit of prey" from a 1903 *Scientific American* article on venomous snakes by Randolph I. Geare; Bingham, "Discovery of Machu Picchu," 712.
11. Hiram Bingham III, "In the Wonderland of Peru,"

National Geographic Magazine 24 (1913): 387; Bingham, field notebook, 41.

12. Bingham, "Discovery of Machu Picchu," 714.
13. Hiram Bingham III to Alfreda M. Bingham, July 26, 1911, MS 81, box 15, folder 38, Bingham Family Papers; Richard S. Lull, "Glacial Man," *Yale Review,* April 1912, 376–89.
14. Bingham, *Portrait,* 147.
15. George F. Eaton, "The Collection of Osteological Material from Machu Picchu," *Memoirs of the Connecticut Academy of Arts and Sciences* 5 (1916): 24.
16. Editor's note, in Bingham, "Wonderland of Peru," 387.
17. Richard L. Burger, in-person interview, January 22, 2015.
18. Lucy C. Salazar, "Machu Picchu: Mysterious Royal Estate in the Cloud Forest," in Burger and Salazar, *Machu Picchu,* 27.
19. Bingham, *Portrait,* 354; John H. Rowe, "Machu Picchu a la luz de documentos del siglo XVI," *Histórica* 14 (1990): 139–54; Salazar, "Mysterious Royal Estate," 27.
20. Arthur Lubow, "The Possessed," *New York Times,* June 24, 2007, http://www.nytimes.com/2007/06/24/magazine/24MachuPicchu-t.html.
21. Bingham, *Portrait,* 310n.
22. Lubow, "Possessed."
23. Ibid.; Bethany L. Turner, "The Servants of Machu Picchu: Life Histories and Population Dynamics in Late Horizon Peru" (Ph.D. diss., Emory University, 2008), 7.
24. Amy C. Hall, "Collecting a 'Lost City' for Science: Huaquero Vision and the Yale Peruvian Expeditions to Machu Picchu, 1911, 1912, and 1914–15," *Ethnohistory* 59 (2012), 304, 294.
25. Lucy C. Salazar and Richard L. Burger, "La historiografía de Hiram Bingham III: Tempestad en los Andes" (talk, International Conference of the Society of American Archaeology, Lima, Peru, August 9, 2014).
26. Ibid.
27. Yale University, "Statement from Yale University Regarding Machu Picchu Archaeological Materials" (press release, November 21, 2010).

Chapter 18: Teaching Evolution

1. Henry F. Osborn, "The Origin of Species, 1859–1925," in "Occasion of the Dedication of the New Museum," *Peabody Bulletin,* 25–38.
2. "Evolution Museum Dedicated at Yale," *New York Times,* December 30, 1925, 9.
3. Schuchert, "Rise of Natural History Museums," 23.
4. Charles Schuchert to John V. Farwell, November 28, 1919, Records of the Trustees, vol. 2 (1917–29), 38, Yale Peabody Museum of Natural History Archives; Charles Schuchert to George MacCurdy, June 14, 1922, MS 435, box 37, pp. 690–91, Charles Schuchert Papers, Manuscripts and Archives, Yale University Library.
5. Schuchert, "Rise of Natural History Museums," 22; Records of the Trustees, vol. 2, 110; Schuchert to Farwell, November 28, 1919, 39, 42–43; Schuchert to MacCurdy, June 14, 1922.
6. Copy of letter from Charles Schuchert to James R. Angell, May 18, 1921, Records of the Trustees, vol. 2, 89–90.
7. John A. Farrell, *Clarence Darrow: Attorney for the Damned* (New York: Vintage Books, 2011), 366.
8. "Whence Man?" *Time,* June 1, 1925, 16.
9. Minutes of the Executive Committee, May 4, 1922, Records of the Trustees, vol. 2, 115; James R. Angell to Charles Schuchert, June 23, 1922, Records of the Trustees, vol. 2, 126.
10. George G. Simpson, "Memorial to Richard Swann Lull (1867–1957)," *Proceedings Volume of the Geological Society of America: Annual Report for 1957* (1958): 128; Harold Callender, "As Modern Science Sums Up the Case for Evolution," *New York Times,* July 19, 1925, BR7.
11. Richard S. Lull, *The Ways of Life* (New York: Harper & Brothers, 1925), 338; telegram from Clarence Darrow to Richard S. Lull, July 10, 1925, Vertebrate Paleontology Archives, Yale University Peabody Museum of Natural History.
12. Winterton C. Curtis, "A Defense Expert's Impressions of the Scopes Trial," http://law2.umkc.edu/faculty/projects/ftrials/scopes/wccurtisaccount.html.
13. "Is Darrow an Infidel or Not? Dayton's Query,"

Chicago Daily Tribune, May 27, 1925, 3; Sam Roberts, "80 Years Ago, They Inherited the Wind," *New York Times,* July 26, 2005, http://www.nytimes.com/2005/07/26/science/80-years-ago-they-inherited-the-wind.html.

14. H. L. Mencken, "Mencken Declares Strictly Fair Trial Is Beyond Ken of Tennessee Fundamentalists," *Baltimore Evening Sun,* July 16, 1925, and H. L. Mencken, "Malone the Victor, Even Though Court Sides with Opponents, Says Mencken," *Baltimore Evening Sun,* July 18, 1925, both reprinted in *H. L. Mencken on Religion,* ed. S. T. Joshi (Amherst, N.Y.: Prometheus Books, 2002), 197, 201; Marion E. Rodgers, *Mencken: The American Iconoclast* (New York: Oxford University Press, 2005), 292.
15. Randy Moore, "The Lingering Impact of the Scopes Trial on High School Biology Textbooks," *BioScience* 51 (2001): 791–92; George M. Price, "Bringing Home the Bacon," *Bible Champion* 35 (1929): 205.
16. "Yale Moves Fossils to Peabody Museum," *New York Times,* March 13, 1925, 5; *General Guide to the Exhibition Halls of the Peabody Museum of Natural History, Yale University* (New Haven, Conn.: [Peabody Museum of Natural History], 1927), 6; notebook of Carl O. Dunbar, June 20, 1917, Carl Dunbar Archives, Invertebrate Paleontology Archives, Yale University Peabody Museum of Natural History.
17. *General Guide to the Exhibition Halls,* 6.
18. "Propaganda at Yale," *New York Telegram-Mail,* March 14, 1925, in newspaper scrapbook labeled "Scrapbook PM Dedication, 1925–1926," Yale University Peabody Museum of Natural History Archives; Arthur Brisbane, "Today" (syndicated column appearing in the *Denver Post* and the *Pittsburgh Press,* among other outlets), December 23, 1925.
19. Anna Kuchment, "Museums Tiptoe Around Climate Change," *Dallas Morning News,* June 14, 2015, http://www.dallasnews.com/news/local-news/20140614-museums-tiptoe-around-climate-change.ece; newspaper clipping, *Springfield* (Mass.) *Republican,* December 26, 1925; "Propaganda at Yale," *New York Telegram-Mail,* newspaper scrapbook, Yale University Peabody Museum of Natural History Archives.
20. James R. Angell to Richard S. Lull, March 5, 1923, Richard Swann Lull Papers, Vertebrate Paleontology Archives, Yale University Peabody Museum of Natural History; "Yale Moves Fossils," *New York Times;* minutes of the annual meeting of the Peabody Museum Trustees, November 24, 1924, Records of the Trustees, vol. 2, 251; "Yale to Co-operate with Local Schools," *New York Times,* October 13, 1925, 2.
21. "Resumé of Report," 1927, Records of the Trustees of the Peabody Museum, Yale Peabody Museum of Natural History Archives.
22. Richard S. Lull, "The New Peabody Museum, Part I: Building and Equipment," *Museum Work* 7 (1924–25): 110.
23. Mildred C. B. Porter, "Behavior of the Average Visitor in the Peabody Museum of Natural History, Yale University," *American Association of Museums,* n.s., no. 16 (1938): 15.

SIDEBAR: THINKING WITH THE SEAT OF ITS PANTS?

a. O. C. Marsh, "Principal Characters of American Jurassic Dinosaurs, Part IV: Spinal Cord, Pelvis, and Limbs of Stegosaurus," *American Journal of Science* 22 (1881): 167; Bert L. Taylor, "The Dinosaur," *Chicago Daily Tribune,* February 26, 1903, 6; David B. Williams, "Dino Brains and Poetry," *GeologyWriter.com* (blog), January 23, 2013, http://geologywriter.com/blog/stories-in-stone-blog/dino-brains-and-poetry.
b. Taylor, "The Dinosaur"; "Mounting Skeleton of 70 Foot Dinosaur at Field Museum: It Was So Big That It Required Two Brains to Move It About," *Chicago Daily Tribune,* February 25, 1903, 5.
c. Bert L. Taylor, "The Dinosaur," in *A Line-o'-Verse or Two* (Chicago: Reilly & Britton, 1911), 75.

SIDEBAR: EVOLUTION AND OTHER PEOPLE'S MORALS

a. George G. Simpson, *Concession to the Improbable: An Unconventional Autobiography* (New Haven, Conn.: Yale University Press, 1978), 38.

b. Léo F. Laporte, “Travel as a Predictor of Scientific Innovation: The Corroborating Case of George G. Simpson,” *Proceedings of the California Academy of Sciences* 55, supplement 2 (2004): 144.

Chapter 19: The Rise of Modern Ecology

1. Edward O. Wilson, foreword to Nancy G. Slack, *G. Evelyn Hutchinson and the Invention of Modern Ecology* (New Haven, Conn.: Yale University Press, 2010), ix.
2. G. Evelyn Hutchinson, *The Kindly Fruits of the Earth: Recollections of an Embryo Ecologist* (New Haven, Conn.: Yale University Press, 1979), 11, 19.
3. Ibid., 237
4. Ibid., 235–36.
5. Tom Lovejoy, “Reflections” (talk, *Journey of the Universe* film premiere, Yale University Peabody Museum of Natural History, New Haven, Conn., March 26, 2011); G. Evelyn Hutchinson, *The Enchanted Voyage* (New Haven, Conn.: Yale University Press, 1962), 97.
6. Hellmut deTerra to James R. Angell, March 22, 1932, MS 263, box 13, folder 951, Ross Granville Harrison Papers, Manuscripts and Archives, Yale University Library.
7. G. Evelyn Hutchinson, *The Clear Mirror* (London: Cambridge University Press, 1936), 112–13; Simone Weil, epigraph to G. Evelyn Hutchinson, *The Itinerant Ivory Tower: Scientific and Literary Essays* (New Haven, Conn.: Yale University Press, 1953). The original epigraph is in French and reads: *La vraie definition de la science, c'est qu'elle est l'étude de la beauté du monde.* Sharon E. Kingsland, “The Beauty of the World: Evelyn Hutchinson's Vision of Science,” in *The Art of Ecology*, ed. David K. Skelly, David M. Post, and Melinda D. Smith (New Haven, Conn.: Yale University Press, 2010), 1.
8. Gordon A. Riley, “Reminiscences of an Oceanographer,” ca. 1982–83, 7–8, MS in the possession of Nancy G. Slack.
9. G. Evelyn Hutchinson, “Concluding Remarks,” *Cold Spring Harbor Symposia on Quantitative Biology* 22 (1957): 416.
10. Melinda D. Smith and David K. Skelly, “Reflection Thereon: G. Evelyn Hutchinson and Ecological Theory,” in Skelly, Post, and Smith, *Art of Ecology*, 159.
11. Stephen J. Gould, *Wonderful Life: The Burgess Shale and the Nature of History* (New York: Norton, 1989), 77; Thomas E. Lovejoy, foreword to Skelly, Post, and Smith, *The Art of Ecology*, x.
12. Martin Kent, “Classics in Physical Geography Revisited,” *Progress in Physical Geology* 24 (2000): 258–59.
13. Lovejoy, foreword, x.

Chapter 20: The Beauty of the Beasts

1. Susan Schlee, “The R/V Atlantis and Her First Oceanographic Institution,” in *Oceanography: The Past*, ed. Mary Sears and Daniel Merriman (New York: Springer-Verlag, 1980), 53–54.
2. Karen A. Rader and Victoria E. M. Cain, *Life on Display: Revolutionizing U.S. Museums of Science and Natural History in the Twentieth Century* (Chicago: University of Chicago Press, 2014), 164; Stephen T. Asma, “Dinosaurs on the Ark: The Creation Museum,” *Chronicle of Higher Education*, May 18, 2007.
3. Rader and Cain, *Life on Display*, 98.
4. Carl O. Dunbar, “Recollections on the Renaissance of Peabody Museum Exhibits, 1939–1959,” *Discovery* 12, no. 1 (1976): 17–19.
5. G. Evelyn Hutchinson, *The Ecological Theater and the Evolutionary Play* (New Haven, Conn.: Yale University Press, 1965), 99–100.
6. Rudolph F. Zallinger, “Creating the Mural,” in *The Age of Reptiles: The Art and Science of Rudolph Zallinger's Great Dinosaur Mural at Yale*, ed. Rosemary Volpe (New Haven, Conn.: Peabody Museum of Natural History, 2007), 7; Lee Grimes, “An Interview with Rudolph F. Zallinger,” *Discovery* 11, no. 1 (1975): 34.
7. Rudolph F. Zallinger, in discussion with John Ostrom, Louise DeMars, Zelda Edelson, Sara Martin, and Copeland MacClintock, January 31, 1980, Rudolph F. Zallinger biographical file, Yale University Peabody Museum of Natural History Archives.
8. Zallinger, “Creating the Mural,” 8.
9. Zallinger, discussion; Wally Swist, “A Salute to the

Jr. and Jerome A. Jackson (Cambridge, Mass.: Nuttal Ornithological Club, 1995), 99; "New Haven Party," *New Yorker,* November 14, 1959, 43–44.

4. Geoffrey T. Hellman, "Curator Getting Around," *New Yorker,* August 26, 1950, 33, 36; Harris, *Capital Culture,* 101.
5. Robin W. Winks, *Cloak and Gown: Scholars in the Secret War* (New York: William Morrow, 1987), 64.
6. Hellman, "Curator Getting Around," 32–33.
7. Ibid., 44; Stanley C. Ball, "Annual Report for 1945–46," RU 471, accession 19ND-A-146, box 14, folder 24, Peabody Museum of Natural History, Yale University, records, Manuscripts and Archives, Yale University Library.
8. Bruce M. Beehler, Roger F. Pasquier, and Warren B. King, "In Memoriam: S. Dillon Ripley, 1913–2001," *Auk* 119 (2002): 1110.
9. Hellman, "Curator Getting Around"; Carl O. Dunbar to S. Dillon Ripley, September 6, 1950, Records of the Director, Yale Peabody Museum of Natural History Archives.
10. "Ripley Finds Rare Bird in Remote Nepal," newspaper clipping in S. Dillon Ripley biographical file, Yale Peabody Museum of Natural History Archives.
11. Alex Dornburg et al., "A Survey of the Yale Peabody Museum Collection of Egyptian Mammals Collected During Construction of the Aswan High Dam, with an Emphasis on Material from the 1962–1965 Yale University Prehistoric Expedition to Nubia," *Bulletin of the Peabody Museum of Natural History* 52 (2011): 266; "Smithsonian Boss Is Former Secret Agent," *Sarasota Herald-Tribune,* March 13, 1964, 17.
12. S. Dillon Ripley, transcript of speech on acceptance of Verrill Medal, January 27, 1984, S. Dillon Ripley biographical file, Yale Peabody Museum of Natural History Archives.
13. Joseph S. Fruton, *Eighty Years* (New Haven, Conn.: Epikouros, 1994), 142–45; Elizabeth J. Rosenthal, *Birdwatcher: The Life of Roger Tory Peterson* (Guilford, Conn.: Lyons, 2008), 233; Martha Freeman, ed., *Always, Rachel: The Letters of Rachel Carson and Dorothy Freeman, 1952–1964* (Boston: Beacon, 1994), 394.
14. Harris, *Capital Culture,* 106.
15. Ibid., 107; G. Evelyn Hutchinson to Rebecca West, August 11, 1963, Collection 1986-002, box 24, folder 8, Rebecca West Papers, Department of Special Collections and University Archives, McFarlin Library, University of Tulsa.
16. Keith S. Thomson, phone interview, April 16, 2015.
17. Ibid.
18. Clement Markert, paraphrased in Slack, *G. Evelyn Hutchinson,* 389; Thomson, interview.
19. Edward O. Wilson, *Naturalist* (Washington, D.C.: Island, 1994), 220.
20. Richard Schodde, "Charles G. Sibley, 1911–1998," *Emu* 100 (2000): 75.
21. Frederick Sheldon, phone interview, April 6, 2015.
22. Tim Birkhead, Jo Wimpenny, and Bob Montgomerie, *Ten Thousand Birds: Ornithology since Darwin* (Princeton, N.J.: Princeton University Press, 2014), 110.
23. Henry Chauncey Jr., personal communication, April 15, 2015.
24. Clive Gammon, "The Great Egg Robbery," *Sunday Times* (London), June 16, 1974, 17.
25. Ibid., 19.
26. "For the Birds," *New York Times,* July 17, 1974, 36; Gammon, "The Great Egg Robbery," 19; Keith Thomson, personal communication, April 6, 2015.

SIDEBAR: "REMINGTON IS RIGHT"

a. Willard D. Hartman to Charles L. Remington, quoting Karl Waage, February 5, 1959, Charles Lee Remington Archives, Entomology, Yale Peabody Museum, New Haven, Conn.
b. Robert A. Raguso, personal communication, June 19, 2014; Stan P. Rachootin, personal communication, March 16, 2015; Robert M. Pyle, *Walking the High Ridge: Life as Field Trip* (Minneapolis: Milkweed, 2000), 45–46.
c. Rachootin, personal communication.
d. John G. Coutsis, quoted in Debra Piot, "Conservation and Collaboration: The Transformative Activism of Charles Lee Remington" (Ph.D. diss., Union Institute and University, 2011), 510.
e. Raguso, personal communication.
f. Stan Rachootin, quoted in Piot, "Conservation and Collaboration," 481.

g. Raguso, personal communication.
h. Pyle, *High Ridge,* 46.
i. Alfred W. Crompton, phone interview, April 2015.
j. Lincoln P. Brower, quoted in Piot, "Conservation and Collaboration," 384.
k. Lincoln P. Brower, Jane Van Zandt, and Joseph M. Corvino, "Plant Poisons in a Terrestrial Food Chain," *Proceedings of the National Academy of Sciences* 57 (1967): 893–98.
l. Robert M. Pyle, "The Origin and History of Insect Conservation in the United States," in *Insect Conservation: Past, Present and Prospects,* ed. Tim R. New (Dordrecht, the Netherlands: Springer, 2012), 163.
m. Lincoln P. Brower, personal communication, September 5, 2014.

SIDEBAR: GRACE UNDER PRESSURE

a. Henry Chauncey Jr., personal communication, April 15, 2015.
b. Anna M. Bidder and Penelope Jenkin, "Grace Evelyn Pickford, 1902–1986," *Newnham College Roll Letter* (1987), 75; Donna Haraway, personal communication, March 25, 2015.
c. Bidder and Jenkin, "Pickford," 76.
d. Dennis J. Taylor, in-person interview, March 25, 2015.

Chapter 24: The Man Who Saved Dinosaurs

1. Charles R. Darwin to James D. Dana, January 7, 1863, Darwin Correspondence Database, https://www.darwinproject.ac.uk/letter/entry-3905.
2. John H. Ostrom, "Osteology of *Deinonychus antirrhopus,* an Unusual Theropod from the Lower Cretaceous of Montana," *Peabody Museum of Natural History Bulletin* 30 (1969): 12–162.
3. John N. Wilford, *The Riddle of the Dinosaur* (New York: Knopf, 1985), 86.
4. Peter Dodson, phone interview, October 2, 2013.
5. Brinkman, personal communication.
6. *Morosaurus* is now called *Camarasaurus.* O. C. Marsh, "Principal Characters of American Jurassic Dinosaurs, Part III," *American Journal of Science* 19 (1880): 255; O. C. Marsh, "Principal Characters of American Jurassic Dinosaurs, Part VI: Restoration of *Brontosaurus,*" *American Journal of Science* 26 (1883): 82; Robert T. Bakker, "Dinosaur Renaissance," *Scientific American,* April 1975, 58.
7. Ostrom, "Osteology of *Deinonychus antirrhopus,*" 139.
8. Bakker, "Dinosaur Renaissance."
9. Peter Dodson and Philip Gingerich, introduction to "Functional Morphology and Evolution," *American Journal of Science* 293-A (1993): vii; Wilford, *Riddle of the Dinosaur,* 174.
10. John H. Ostrom, "Terrestrial Vertebrates as Indicators of Mesozoic Climates," in *Proceedings of the North American Paleontological Convention: Field Museum of Natural History, Chicago, September 5–7, 1969,* vol. 1, part D, ed. Ellis L. Yochelson (Lawrence, Kans.: Allen, 1970), 360; Robert Bakker, interview by Robert Siegel, in "Influential Paleontologist John Ostrom, 77, Dies," *All Things Considered,* NPR, July 21, 2005.
11. Edward Hitchcock, handwritten draft of poem, reproduced in S. George Pemberton, "History of Ichnology: Early Ichnology Poems and Their Poets," *Ichnos* 17 (2010): 268.
12. John H. Ostrom, "The Origin of Birds," *Annual Review of Earth and Planetary Sciences* 3 (1975): 62; Daniel Brinkman, in-person interview, October 2, 2013.
13. Richard O. Prum et al., "A Comprehensive Phylogeny of Birds (Aves) Using Targeted Next-Generation DNA Sequencing." *Nature* 526 (2015): 569–73.
14. Wilford, *Riddle of the Dinosaur,* 88.
15. Stephen Jay Gould, "Dinomania," *New York Review of Books,* August 12, 1993, http://www.nybooks.com/articles/archives/1993/aug/12/dinomania/.
16. John N. Wilford, "John H. Ostrom, Influential Paleontologist, Is Dead at 77," *New York Times,* July 21, 2005.
17. "John Ostrom," *Times* (London), August 9, 2005, http://www.thetimes.co.uk/tto/opinion/obituaries/article2084182.ece; Bakker, NPR interview.
18. John H. Ostrom, "How Bird Flight Might Have Come About," in *Dinofest International: Proceedings of a Symposium Held at Arizona State University,* ed.

Donald L. Wolberg, Edmund Stump, and Gary D. Rosenberg (Philadelphia: Academy of Natural Sciences, 1997), 309.

19. Brinkman, interview.

Epilogue

1. Franklin Parker and Betty J. Parker, "George Peabody (1795–1869), Education: A Debt Due from Present to Future Generations (June 16, 1852)," ERIC Number ED474157, Washington, D.C.: ERIC Clearinghouse (2002), http://eric.ed.gov/?id=ED474157.
2. Jakob Vinther, "Fossil Colors," http://www.jakobvinther.com/Fossil_color.html; Derek E. G. Briggs, "Adolf Seilacher, (1925–2014)," *Nature* 509 (2014): 428.
3. Jakob Vinther, personal communication, April 22, 2015.
4. Ibid.
5. Quanguo Li et al., "Plumage Color Patterns of an Extinct Dinosaur," *Science* 327 (2010): 1369–72.
6. Kirk Johnson, "The Future of Natural History Museums" (talk, Yale Peabody Museum of Natural History, New Haven, Conn., April 22, 2015).
7. Wilmarth Lewis, *The Yale Collections* (New Haven, Conn.: Yale University Press, 1946), ix–x.
8. Johnson, "Future of Natural History Museums."
9. Michael J. Donoghue and William S. Alverson, "A New Age of Discovery," *Annals of the Missouri Botanical Garden* 87 (2000): 110–26.
10. E.g., Jane J. Lee, "Pictures: Top 10 Newly Discovered Species of 2012," *National Geographic News,* May 24, 2013, http://news.nationalgeographic.com/news/2013/05/pictures/130523-top-ten-new-species-animals-plants-science/; Jonathan Jones, "If the Lesula is a Newly Discovered Monkey, Why Is It So Oddly Familiar?" *Guardian,* September 13, 2012, http://www.theguardian.com/commentisfree/2012/sep/13/lesula-new-monkey-familiar; John Hart, quoted in Andrea Mustain, "New, Colorful Monkey Species Discovered," *LiveScience,* September 12, 2012, http://www.livescience.com/23147-monkey-species-discovered.html.
11. Eric J. Sargis, quoted in "In Congo, a Rare Find—A New Species of Monkey," *YaleNews,* September 12, 2012, http://news.yale.edu/2012/09/12/congo-rare-find-new-species-monkey.

ILLUSTRATION CREDITS

Page ii Division of Vertebrate Paleontology Archives [YPM VPAR 000240]; restoration of the skeleton of *Ichthyornis dispar* Marsh, from *Odontornithes: A Monograph on the Extinct Toothed Birds of North America,* by O. C. Marsh [1880]. U.S. Geological Exploration of the Fortieth Parallel. Government Printing Office: Washington, D.C., pl. 26; photography by William K. Sacco.

Page vi Division of Invertebrate Paleontology [YPM IP 006657], Yale Peabody Museum.

Page ix Division of Vertebrate Paleontology [YPM VP 001831, YPM VP 005775, YPM VPAR 000097], Yale Peabody Museum

Page 3 Division of Vertebrate Paleontology Archives [YPM VPAR 000072], Yale Peabody Museum

Page 4 "The Yale College Expedition of 1870," *Harper's New Monthly Magazine,* no. 257, October 1871, 667

Page 7 Division of Anthropology [YPM ANT 001982], Yale Peabody Museum

Page 10 Oil on ivory miniature by Nathaniel Rogers, c. 1815; Yale University Art Gallery, Gift of Miss Maria Trumbull Dana, 1954.34.1

Page 11 Division of Historical Scientific Instruments [YPM HSI 020001], Yale Peabody Museum; photography by Robert Lorenz

Page 15 Division of Mineralogy and Meteoritics [YPM MIN 100375], Yale Peabody Museum; Benjamin Silliman and James Luce Kingsley, "An Account of the Meteor, which burst over Weston in Connecticut, in December, 1807, and of the falling of stones on that occasion." *Memoirs of the Connecticut Academy of Arts and Sciences* vol. 1, issue 1, no. 15, 141–161 [1810]

Page 16 Yale Peabody Museum Archives; photography by Robert Lorenz

Page 17 Yale Peabody Museum Archives; Division of Mineralogy and Meteoritics [YPM MIN 015349, YPM MIN 051344, YPM MIN 051786], Yale Peabody Museum; photography by Robert Lorenz

Page 20 Division of Mineralogy and Meteoritics Archives, Yale Peabody Museum; photography by Robert Lorenz

Page 23 Division of Vertebrate Paleontology Archives [YPM VPAR 000038], Yale Peabody Museum

Page 30 Division of Vertebrate Paleontology [YPM VPR 00309, YPM VP 001648], Yale Peabody Museum; photography by Robert Lorenz

Page 33 Division of Invertebrate Paleontology [YPM IP 054701, YPM IP 009205, YPM IP 009755, YPM IP 006467, YPM IP 001853], Yale Peabody Museum; photography by Robert Lorenz

Page 37 Oil on canvas by Daniel Huntington, 1867, Yale University Art Gallery, 1867.1

Page 40 Oil on canvas by Daniel Huntington, 1858, Yale University Art Gallery, Bequest of Edward Salisbury Dana, B.A. 1870, 1961.46

Page 43 "Wreck of the Peacock," drawing by Alfred T. Agate, 1841, Navy Art Collection, Archives Branch, Naval History and Heritage Command, Washington, D.C., catalog no. NH 51496

Page 45 Division of Invertebrate Zoology [YPM IZAR 005905, YPM IZ 001999], Yale Peabody Museum; photography by Robert Lorenz

Page 47 Division of Mineralogy and Meteoritics Archives [YPM MINAR 000076, YPM MINAR 000112], Yale Peabody Museum; photography by Robert Lorenz

Page 52 Division of Vertebrate Paleontology [YPM VP 000312], Yale Peabody Museum; photography by Robert Lorenz

Page 54 Division of Vertebrate Paleontology [YPM VP 058534], Yale Peabody Museum; photography by Robert Lorenz

Page 56 E. D. Cope, "Synopsis of the Extinct Batrachia, Reptilia and Aves of North America. Part I," *Transactions of the American Philosophical Society* 14 [1869, Revised 1870]

Page 61 Yale Peabody Museum Archives [YPMAR 001010]

Page 65 Yale Peabody Museum Archives [YPMAR 001012]

Page 66 Division of Vertebrate Paleontology Archives [YPM VPAR 000240], Yale Peabody Museum; restoration of the skeleton of *Ichthyornis dispar* Marsh, from O. C. Marsh, *Odontornithes: A Monograph on the Extinct Toothed Birds of North America* [1880], U.S. Geological Exploration of the Fortieth Parallel, Government Printing Office: Washington, D.C., pl. 26; photography by William K. Sacco

Page 68 Wellcome Library, London, no. 12956i

Page 69 Division of Vertebrate Paleontology [YPM VP 011340], Yale Peabody Museum; photography by Robert Lorenz

Page 71 Division of Vertebrate Paleontology Archives [YPM VPAR 002503, YPM VPR 00372], Yale Peabody Museum; photography by Robert Lorenz

Page 74 Division of Anthropology [YPM ANT 049270], Yale Peabody Museum; photography by William K. Sacco

Page 75 (top) Division of Vertebrate Paleontology [YPM VP 001778], Yale Peabody Museum

Page 75 (bottom) Division of Vertebrate Paleontology Archives [YPM VPAR 002540], Yale Peabody Museum

Page 81 Yale Peabody Museum Archives [YPMAR 002586]

Page 83 Division of Vertebrate Paleontology Archives [YPM VPAR 000096], Yale Peabody Museum

Page 84–85 Division of Vertebrate Paleontology Archives [YPM VPAR 002241], Yale Peabody Museum; art by Arthur Lakes, 1879

Page 86 Division of Vertebrate Paleontology [YPM VP 004835, YPM VPAR 000291], Yale Peabody Museum; photography by Robert Lorenz

Page 89 Division of Vertebrate Paleontology Archives [YPM VPAR 002405], Yale Peabody Museum; art by Arthur Lakes, 1879

Page 96 Images of Yale individuals, ca. 1750–2001 [inclusive], Manuscripts and Archives, Yale University, image no. 827

Page 98 Catlin's North American Indian Portfolio, pl. 6, ZZc12 844ca. Yale Collection of Western Americana, Beinecke Rare Book and Manuscript Library, Yale University; art by George Catlin

Page 103 Yale Peabody Museum Archives; photography by Jerry Domian

Page 106 Division of Vertebrate Paleontology Archives [YPM VPAR 002473], Yale Peabody Museum

Page 107 Division of Vertebrate Paleontology Archives [YPM VPAR 000186], Yale Peabody Museum
Page 108 Division of Vertebrate Paleontology Archives [YPM VPAR 000345], Yale Peabody Museum
Page 109 Yale Peabody Museum Archives [YPMAR 002036]
Page 112 Yale Peabody Museum Archives [YPMAR 001103B]
Page 114 Division of Invertebrate Zoology [YPM IZ 010272 GP, YPM IZ 009634 GP], Yale Peabody Museum; photography by Robert Lorenz
Page 115 Division of Invertebrate Zoology, Yale Peabody Museum, from A. E. Verrill, "The Cephalopods of the North-eastern coast of America, Part I: The Gigantic Squids (*Architeuthis*) . . . ," *Transactions of the Connecticut Academy of Sciences* 5 [1879]. Yale Peabody Museum Archives.
Page 117 Images of Yale individuals, ca. 1750–2001 [inclusive], Manuscripts and Archives, Yale University, image no. 8121
Page 119 From Daniel Cady Eaton, *The Ferns of North America*. Drawings by J. H. Emerton and C. E. Faxon. Salem and Boston: S. E. Cassino, 1879, vol. 1 (pl. XXXIX), Yale University Library, QK525. E3.1; photography by Robert Lorenz. On permanent loan to the Yale Herbarium, Division of Botany Archives, Yale Peabody Museum
Page 120 Smithsonian Institution Archives, image no. 78-10629
Page 128 Yale Peabody Museum Archives [YPMAR 000592]
Page 130 Division of Vertebrate Paleontology [YPM VP 000255], Yale Peabody Museum; photography by Robert Lorenz
Page 132 Courtesy of the Department of Paleobiology, National Museum of Natural History, Smithsonian Institution
Page 138 (left) Yale Peabody Museum Archives, oil portrait of O. C. Marsh by Thomas LeClear, date unknown
Page 138 (right) Division of Vertebrate Paleontology Archives [YPM VPAR 002544]
Page 147 Division of Invertebrate Paleontology [YPM IP 038278, YPM IP 202147, YPM IPAR 000814, YPM IP 000228, YPM IPAR 000549], Yale Peabody Museum; photography by Robert Lorenz
Page 149 Division of Invertebrate Paleontology Archives [YPM IPAR 000815], Yale Peabody Museum
Page 151 George Reber Wieland Papers (MS 750). Manuscripts and Archives, Yale University Library
Page 153 Division of Paleobotany [YPM PB 151264, YPM PB 151288, YPM PB 005069, YPM PB 000055], Yale Peabody Museum; photography by Robert Lorenz
Page 156 Division of Vertebrate Paleontology [YPM VP 002182, YPM VPAR 000174], Yale Peabody Museum
Page 160 Division of Invertebrate Paleontology [*clockwise, from top center:* YPM IP 204444 (*Lingula*), YPM IP 527825, 527829, 527830, 527833 (*Rafinesquina*), YPM IP 006231 (*Pentamerus*), YPM IP 527677, 527678 (*Rhynchopora*), YPM IP 527798, 527799, 527801 (*Athyris*), YPM IPS 003187 (*Mucrospirifer*), YPM IPS 003690 (*Terebratula*), YPM IP 007526 (*Hesperorthis*), YPM IP 055628 (*Terebratella*), YPM IP 035744 (*Pygope*), YPM IP 527676 (*Magellania*), YPM IPS 002884 (*Derbyia*), YPM IP 204437 (*Hercosia*), YPM IP 509316 (*Gigantella*)], Yale Peabody Museum; photography by Robert Lorenz
Page 164 Division of Invertebrate Paleontology [YPM IP 000001], Yale Peabody Museum; photography by Robert Lorenz
Page 167 Division of Invertebrate Paleontology Archives [YPM IPAR 002489], Yale Peabody Museum
Page 173 Yale University / Hiram Bingham / National Geographic Creative
Page 176 Yale University / Hiram Bingham / National Geographic Creative

Page 180 Division of Anthropology [YPM ANT 017973], Yale Peabody Museum; photography by William K. Sacco

Page 185 Yale Peabody Museum Archives [YPMAR 002432]

Page 187 Yale Peabody Museum Archives [YPMAR 002010]

Page 189 From *Time Magazine,* June 1, 1925 © 1925 Time Inc. Used under license

Page 190 Division of Vertebrate Paleontology Archives [YPM VPAR 000280], Yale Peabody Museum; Yale Peabody Museum Archives [YPMAR 002198]

Page 193 Division of Vertebrate Paleontology Archives, Yale Peabody Museum; restoration of *Stegosaurus ungulatus* Marsh, pl. 52 from O. C. Marsh, *The Dinosaurs of North America,* Sixteenth Annual Report of the U.S. Geological Survey, Washington, D.C.; photography by William K. Sacco

Page 195 Yale Peabody Museum Archives [YPMAR 002190]

Page 197 Yale Peabody Museum Archives [YPMAR 002031]; Division of Vertebrate Paleontology [YPM VP 001980], Yale Peabody Museum

Page 200 Division of Entomology [YPM ENT 451310], Yale Peabody Museum; photography by Robert Lorenz

Page 201 G. Evelyn Hutchinson papers, 1875–1992 [inclusive], 1922–1991 [bulk], Manuscripts and Archives, Yale University, image no. 6290

Page 206 Division of Entomology Archives [YPM ENTAR 001733], Yale Peabody Museum; photography by Robert Lorenz

Page 211 Yale Peabody Museum Archives [YPMAR 000802B]

Page 215 Education Department Archives, Yale Peabody Museum

Page 216 Rudolph Zallinger, courtesy of the Zallinger family. From F. Clark Howell and the Editors of Time-Life Books. 1968. *Early Man.* Life Nature Library. New York: Time, Inc.

Page 221 Yale Peabody Museum Archives; photography by Jerry Domian

Page 223 Yale Peabody Museum Archives [YPMAR 002171I]

Page 227 Division of Anthropology Archives [YPM ANTAR 010956], Yale Peabody Museum

Page 228 Division of Anthropology [YPM ANT 148531, YPM ANT 148975], Yale Peabody Museum; photography by Robert Lorenz

Page 229 Division of Anthropology [YPM ANT 006951, YPM ANTAR 035888], Yale Peabody Museum; photography by Robert Lorenz

Page 231 Division of Anthropology [YPM ANT 057101, YPM ANT 187282, YPM ANT 187304, YPM ANTAR 000251, YPM ANTAR 031289, YPM ANTAR 035886], Yale Peabody Museum; photography by Robert Lorenz

Page 234 Division of Anthropology Archives [YPM ANTAR 035893], Yale Peabody Museum

Page 236 Division of Anthropology Archives [YPM ANTAR 033978], Yale Peabody Museum

Page 237 Division of Anthropology [YPM ANT 264662], Yale Peabody Museum; photography by Robert Lorenz

Page 239 Division of Anthropology Archives [YPM ANTAR 035892], Yale Peabody Museum

Page 240 Division of Anthropology [YPM ANT 230001, YPM ANT 229955, YPM ANT 249192], Yale Peabody Museum; photography by Robert Lorenz

Page 245 Office of Public Affairs, Yale University, photographs of individuals, 1870–2005 [inclusive], Manuscripts and Archives, Yale University; photography by Charles Alburtus, © Yale University, image no. 1064

Page 248 Yale Peabody Museum Archives [YPMAR 000580]

Page 251 Division of Entomology [YPM ENT 451311, YPM ENT 451312], Yale Peabody Museum; photography by Robert Lorenz

Page 259 Division of Invertebrate Zoology Archives [YPM IZAR 005916], Yale Peabody Museum

Page 263 Division of Vertebrate Zoology [YPM ORN 134157], Yale Peabody Museum; photography by Robert Lorenz

Page 267 Division of Vertebrate Paleontology Archives [YPM VPAR 002708], Yale Peabody Museum

Page 269 Courtesy of Karen Ostrom

Page 270 Illustration by Robert Bakker, from John H. Ostrom, "Osteology of *Deinonychus antirrhopus,* an Unusual Theropod from the Lower Cretaceous of Montana," *Bulletin of the Peabody Museum of Natural History* 30 [1969]

Page 272 Illustration © Nobumichi Tamura

Page 276 Illustration by M. DiGiorgio; mdigiorgio.com

Endpapers Detail from *A General View of the Animal Kingdom* wall chart from Anna Maria Redfield, *Zoölogical Science, or Nature in Living Forms.* New York and Hartford, Connecticut: E. B. & E. C. Kellogg, 1858; Division of Invertebrate Paleontology Archives, Yale Peabody Museum

INDEX

Note: Page numbers in *italics* indicate photographs and illustrations.

Labrus, Common Black fish, or N.Y. Tautog.
Ctenolabrus ceruleus
Malthea nasula
Short nosed Malthaea
Labridae. 22 gen. 500 sp.
Wrasses, or Rock fishes.
Pimelodus, Cat fish, Bull pouts, Bull heads, Horn pouts, or Minister.
Galeichthys, Oceanic Catfish.
Cyprinus carpio, Carp. C. auratus, Gold fish. C. gobio, Gudgeon.
Barbus, Barbel.
Labeo, Chubsucker. Fundulus, Killefish. Hydrargira, Minnow.
Belone vulgaris, Garfish. Sea Pike, or Mackerel guide.
Scomber esox, Billfish.
Aulostoma Chinensis, Trumpet fish. Fistularia, Pipe fish.
Fistularidae. 20 sp.
132 sp Salmonidae in fresh and salt water.
Salmo salar, Common Sea Salmon. S. confinis, Lake, or Salmon Trout.
Osmerus, Smelt. Coregonus, White fish, Shad Salmon. S. amethystus
Engraulis, Anchovy. Hyodon, Mooneye. Elops.
C. Sprattus, Sprat. C. sardina, Sardine.
Silurus glanis
Sheat fish.
Esocidae.
Pikes, and Garfish.
Brook Trout.
Clupea, Herring.
C. Pilchardus, Pilchard.
Gadidae. (Codfish Family.) 110 sp.
Morrhua Americana, American Codfish.
Tom Cod, & Haddock. Phycis, Codling.
Gadus morrhua, Common Codfish of Newfoundland.
Cyclopterus Lumpus, Lump
Cock Paddle.
Lota
Merlucius
Echeneis Remora, or Sucking fish.
Anguillidae, or Muraenidae. (Eels.)
Scales so small as to be scarcely apparent in the
(Herrings & Pilchards.)
Pleuronectidae. (Flatfish Family.) 150 species.
Pleuronectes, or
Rhombus maximus,
Turbot.
Platessa,
Flatfish. Dab, and Plaice.
Power Cod,
Flounder,
Rhombus vulgaris, Brill.
Achirus, Sole.
Hippoglossus, Halibut.
Some times 600 lbs, 4 to 8 ft. long.
Anguilla conger, Conger Eel.
Thick as a mans thigh 10 ft. long.
Anguilla vulgaris, Common Eel.
Gymnotus electricus, Electric Eel.
Syngnathidae. (Pipe fishes.)
Syngnathus, Pipe fish, 5 to 16 inches long
Hippocampus, Sea horse.
3 to 6 inches sometimes 10 in. length
The only genus of fish with a prehensile tail.
Acanthosoma, Small Globe fish, armed with spines, one inch long.
Tetraodon hispidus, Globe fish, Puffer, or Blower, Swell fish
Orthagoriscus mola, Short Sun-fish
Phosphorescent at night.
Diodon, Sea Porcupine, Balloon fish
Diodon pilosus
2 to 4 inches
Aluteres
6 to 9 inches
Pegasus.
Planidae, or Pleuronectidae.
Both eyes on the same side,
no air bladder.
Turbot,
Pleuronectes.
Ammodytus, Lance
Ophidium.
Purpura.
Buccinum
Rostellaria,
Chinese Spindle.
Turreted Pleurotoma
Paludina.
Phasianella.
Melania.
Oceanic Snail.
Janthina.
Melanopsis.
Ampullaria.
Egg Ovula.
Ovulum volva
V. Aethiopica
Oliva Textilina,
Figured Olive
Olivella
Episcopalis, Episcopal Mitre.
Terebellum
This long proboscis is white.
Mitra corrugata.
V. undulata, Undulated Volute.
Young shell.
Sigaretus
Cryptostoma
Vermetus.
Magilus
Variegated Triton,
Triton
Turbo marmoratus,
Marbled Turbo.
Turritella,
Screw shell.
Neritina
Natica.
Navicella.
Nerita.
Trochus obeliscus,
Rotella.
Delphinula
Trochus imperialis.
Phorus,
Carrier Trochus.
Physa
Lymnaea
Scalaria, Wendle trap, or Royal
Eulima.
Murex Tenuispina, Venus' comb, or
Strombus.
Conelix.
Columbella
Ovulum
Erato.
Tear shell.
Cancellaria
Marginella